How to Organize and Run a Failure Investigation

Daniel P. Dennies, Ph.D., P.E.

ASM International®
Materials Park, Ohio 44073-0002
www.asminternational.org

Copyright © 2005
by
ASM International®
All rights reserved

No part of this book may be reproduced, stored in a retrieval system, or transmitted, in any form or by any means, electronic, mechanical, photocopying, recording, or otherwise, without the written permission of the copyright owner.

First printing, August 2005

The following case histories are from N. Schlager, Ed., *When Technology Fails: Significant Technological Disasters, Accidents, and Failures of the 20th Century,* Gale Group, 1994, and are used by permission: "BOAC Comet Crashes: Mediterranean Islands of Elba and Stromboli (1957)"; "Tacoma Narrows Bridge Collapse: Washington State (1940)"; "Firestone 500 Steel-Belted Tire Failure (1972–1978)"; "Rocket Fire: Former Soviet Union (1960)"; "Andrea Doria-Stockholm Collision: Off Massachusetts (1957)."

Great care is taken in the compilation and production of this volume, but it should be made clear that NO WARRANTIES, EXPRESS OR IMPLIED, INCLUDING, WITHOUT LIMITATION, WARRANTIES OF MERCHANTABILITY OR FITNESS FOR A PARTICULAR PURPOSE, ARE GIVEN IN CONNECTION WITH THIS PUBLICATION. Although this information is believed to be accurate by ASM, ASM cannot guarantee that favorable results will be obtained from the use of this publication alone. This publication is intended for use by persons having technical skill, at their sole discretion and risk. Since the conditions of product or material use are outside of ASM's control, ASM assumes no liability or obligation in connection with any use of this information. No claim of any kind, whether as to products or information in this publication, and whether or not based on negligence, shall be greater in amount than the purchase price of this product or publication in respect of which damages are claimed. THE REMEDY HEREBY PROVIDED SHALL BE THE EXCLUSIVE AND SOLE REMEDY OF BUYER, AND IN NO EVENT SHALL EITHER PARTY BE LIABLE FOR SPECIAL, INDIRECT OR CONSEQUENTIAL DAMAGES WHETHER OR NOT CAUSED BY OR RESULTING FROM THE NEGLIGENCE OF SUCH PARTY. As with any material, evaluation of the material under end-use conditions prior to specification is essential. Therefore, specific testing under actual conditions is recommended.

Nothing contained in this book shall be construed as a grant of any right of manufacture, sale, use, or reproduction, in connection with any method, process, apparatus, product, composition, or system, whether or not covered by letters patent, copyright, or trademark, and nothing contained in this book shall be construed as a defense against any alleged infringement of letters patent, copyright, or trademark, or as a defense against liability for such infringement.

Comments, criticisms, and suggestions are invited, and should be forwarded to ASM International.

Prepared under the direction of the ASM International Technical Books Committee (2004–2005), Yip-Wah Chung, FASM, Chair.

ASM International staff who worked on this project include Scott Henry, Senior Manager of Product and Service Development; Bonnie Sanders, Manager of Production; and Madrid Tramble, Senior Production Coordinator.

Library of Congress Cataloging-in-Publication Data
Dennies, Daniel P.
How to organize and run a failure investigation / Daniel P. Dennies.
p. cm.
Includes bibliographical references and index.
ISBN: 0-87170-811-6
1. Structural failures—Investigation. I. Title
TA656.D46   2005
624.1′71—dc22   2005045226

ISBN: 0-87170-811-6
SAN: 204-7586

ASM International®
Materials Park, OH 44073-0002
www.asminternational.org

Printed in the United States of America

Contents

Preface v
About the Author vii

CHAPTER 1 What Is a Failure? 1

    Challenges Faced by Failure Analysts 2
    The Failure Analysis Process 3
    Recognizing a Failure 7
    Case History: BOAC Comet Crashes, Mediterranean Island
        of Elba and Stromboli (1957) 20
    Case History Discussion 26

CHAPTER 2 Failures Come in all Shapes and Sizes 29

    Some Failures You Can Touch, and Some You Cannot 30
    What Defines a Failure? 34
    Why Do Failures Happen? 36
    Case History: Tacoma Narrows Bridge Collapse, Washington
        State (1940) 53
    Case History Discussion 59

CHAPTER 3 Aspects of a Failure Investigation 61

    Relevant Statistics 61
    Creating a Database 63
    Why Is a Failure Investigation Performed? 67
    Why Determine Root Cause? 68
    When Is a Failure Investigation Performed? 69
    Benefits of a Failure Investigation 71
    Puzzles and Problem Solving 71
    The Four-Step Problem-Solving Process 76
    Why Plan a Failure Investigation? 77
    Timeline of a Failure Analysis 78

Case History: Firestone 500 Steel-Belted Tire Failure
(1972–1978) 82
Case History Discussion 88

CHAPTER 4 Nine Steps of a Failure Investigation 91

Step 1: Understand and Negotitate the Investigation Goals 91
Step 2: Obtain a Clear Understanding of the Failure 96
Step 3: Objectively and Clearly Identify All Possible
Root Causes 108
Step 4: Objectively Evaluate the Likelihood of Each
Root Cause 113
Step 5: Converge on the Most Likely Root Cause(s) 115
Step 6: Objectively and Clearly Identify All Possible
Corrective Actions 118
Step 7: Objectively Evaluate Each Corrective Action 120
Step 8: Select the Optimal Corrective Action(s) 120
Step 9: Evaluate the Effectiveness of the Selected
Corrective Action(s) 120
Case History: Rocket Fire, Former Soviet Union (1960) 121
Case History Discussion 124

CHAPTER 5 A Few Pitfalls and More Useful Tools 127

Pitfalls 128
Failure Investigation Is Not 131
What Can You Do? 136
Technical Skills 137
Communication Skills 137
Technical Integrity 139
Other Useful Tools 142
Case History: *Andrea Doria-Stockholm* Collision,
Off Massachusetts (1956) 144
Case History Discussion 149

APPENDIX 1: General Procedures for Failure Analysis 151

APPENDIX 2: Glossary 197

APPENDIX 3: Suggested Further Reading 217

Index 225

Preface

This book presents a proven systematic approach and template to advance a failure investigation, including a discussion of the methodology required, organizational tools, and a review of failure investigation concepts. This book provides a learning platform for personnel from all disciplines: materials, design, manufacturing, quality, and management. Guidance is provided in areas such as learning how to define objectives, negotiating the scope of an investigation, examining the physical evidence, and applying general problem-solving techniques.

The systematic approach explained herein examines the relationship between various failure sources (e.g., corrosion) and the organization and conduct of the failure investigation. The examples provided focus on the definition of and requirements for a professionally performed failure analysis of a physical object or structure. However, many of the concepts learned have broader applicability in other areas of business, manufacturing, and life in general.

Professional failure analysis is a multilevel process that includes the metallurgical analysis of the physical part itself, and also much more. This book is intended to showcase some of the latest thinking on how the different "layers" of a failure investigation process should work together leading to a concise, well-supported and well-documented root cause, resulting in corrective action when the investigation is complete.

Failure investigations cross company functional boundaries and are an integral component of any design or manufacturing business operation. Learning the proper steps to organize and professionally conduct investigations is essential for solving manufacturing problems and assisting with redesigns. Examples of how competent materials engineers can use these concepts in a failure investigation are emphasized here.

After completing this book, readers shall have the mindset that a well-organized failure investigation is the proper method and will take action to apply the concepts learned. At first, it may be difficult for your customers to appreciate the time and effort it takes to conduct a successful failure investigation. Perhaps they want you to conduct their investigation a specific way and according to their preferences. They may already have a "silver bullet theory" they just want you to confirm. Remember, Rome was not built in a day. You will not convince anybody in five seconds; it takes time. Once your customers see how well the methods presented in this book work and how convincingly the evidence leads to the root cause and corrective actions, they will come around. Readers shall come to appreciate that a few hours spent in preparation may save a lot of time and money and may even be the key to achieving a successful conclusion to the failure investigation.

ACKNOWLEDGMENTS

I would like to thank the staff at ASM International for their assistance in the completion of this book—in particular, Sarah Fanger, who prompted me to develop the original course, Julie Lorence, whose assistance expanded the course and initiated the book, and Scott Henry, who made the book a reality. I would also like to thank the members of ASM International Failure Analysis Committee, whose members participated in many conversations and stories that fostered the development of the ASM International course and this book. Lastly, the creation of this book would not have been possible without the love and support of my friends and family, especially my parents, Paul and Lillian Dennies.

Daniel P. Dennies, Ph.D., P.E.
Foothill Ranch, California
February 2, 2005

About the Author

Dr. Daniel P. Dennies, FASM, is an Associate Technical Fellow for the Boeing Company and has 26 years of experience as a metallurgist. The majority of his career has been in the U.S. space and aerospace industries working on projects such as the Space Shuttle Main Engine, the National Launch System, the National Aerospace Plane, expendable launch systems like Delta and Titan, and most recently, the International Space Station and Space Shuttle programs. He is an expert in failure analysis and also works as an expert witness. Dr. Dennies is a recipient of the coveted NASA Silver Snoopy Award.

A member of ASM for 28 years, Dr. Dennies was named a Fellow of ASM in 2002. He has held positions on local as well as national committees. He has served as chair of the ASM Chapter Council and has participated as a "Materials Mentor" at Materials Camps sponsored by the ASM Materials Education Foundation since the program began in 2000. Dr. Dennies is currently serving on the Foundation's Board of Trustees. He is also a contributing editor to ASM's *Journal of Failure Analysis and Prevention* and teaches the ASM course, "How to Organize and Run a Failure Investigation." Dr. Dennies received the 2002 ASM Materials Engineering Institute Instructor of Merit Award and the 2004 ASM International Allan Ray Putnam Service Award.

Dr. Dennies has a bachelor's degree in metallurgical engineering from California Polytechnic Institute, San Luis Obispo, a master's degree in materials engineering from the University of Southern California, an MBA from Pepperdine University, and a Ph.D. in material science and engineering from University of California, Davis.

CHAPTER 1

What Is a Failure?

IMPORTANT ASPECTS of failure investigation addressed in this chapter include:

- The complexities of organizing a failure investigation and understanding the importance of defining a clear and concise goal, direction, and plan prior to beginning the investigation
- The many aspects and organizational levels that may contribute to and define a failure
- The role and requirements of leading a successful failure investigation
- The importance of discovering the root cause of a failure through the use of a well-organized and well-planned failure investigation.

Introduction

On January 27, 1967, the crew of Apollo 1 was lost when a fire broke out inside their command capsule during a simulation run aboard a Saturn 1B launch vehicle. This tragic event caused a nation in the midst of the early years of the "space race" to stop, think, and reconsider President John F. Kennedy's May 25, 1961 promise that "we choose to go to the moon in this decade." Weeks later, during the federal investigation of the Apollo 1 fire, astronaut Frank Borman proclaimed it a "failure of imagination." At this early stage of the space program, people simply could not be expected to imagine all the possible problems that could arise.

Two years, five months, and 23 days later, on July 20, 1969, the lunar module *Eagle* landed on the moon. The command capsule, service module, and lunar module used for that historic space flight and landing had been improved by the rigorous and detailed failure investigation of the capsule from the Apollo 1 fire. A second capsule, the next in line, was also sacrificed in order to better understand the failure and determine its root cause. During the investigation, many other design and manufacturing flaws were discovered, with improvements suggested and implemented.

Failure investigation has come a long way since the Apollo 1 fire. Those involved in failure investigation have an incredible array of sophisticated and powerful tools and equipment to assist them. The world of materials science has evolved in many directions, and extensive research has provided more information on materials than any one person could hope to accumulate in a lifetime.

Unfortunately, many failure investigations performed today are not the victims of a failure of imagination, but rather of a failure of organization. Too many failure investigations are undertaken without a clear and concise goal, direction, or plan. Somewhere along the way, the investigation sputters out of control and never achieves its fundamental purpose: discovering root cause.

Using examples from a variety of industries, this book outlines a proven systematic approach to failure investigation and teaches the steps you need to follow. It discusses the effects of various failure sources, such as corrosion, on the organization of the investigation. It provides a learning platform for personnel from all disciplines: materials, design, manufacturing, quality, and management.

Challenges Faced by Failure Analysts

Increasingly, failure investigation is becoming a public forum. As evidenced by recent television shows, our society is interested in criminal investigations and how the process works. More people now read the results of failure investigations and judge—do not just accept—the final answer. In some cases, the conclusions reached by the failure investigation cannot be derived from the evidence presented. It is mandatory that professional failure investigators be prepared and do the best job possible.

For example, consider the public response to the crash investigation of TWA Flight 800. The aircraft in this July 17, 1996 flight carrying 230 people from New York to Paris exploded shortly after takeoff. The National Transportation Safety Board (NTSB) failure investigation took four years, and it concluded that a stray wire in the center wing fuel tank caused ignition of the flammable fuel/air mixture in the tank. However, eight years after the crash, independent investigations are still ongoing that dispute this conclusion. Some of these investigators contend that a shoulder-fired missile hit the plane and that the government has turned a blind eye to certain facts. Regardless of the credibility of these claims, they indicate that high-profile failure investigations are under increased scrutiny today. People do not blindly believe what the experts tell them. Failure analysis professionals must understand that their investigations may be received with skepticism, even if all evidence is considered.

But where to begin? Have you ever been handed a failure investigation and felt unsure of all the steps required to complete it? Or perhaps you had to review a failure investigation and wondered if all the aspects had been properly covered? After reading the results of a failure investigation, do you know what to do next? The initial steps of a failure investigation set the direction and, in many cases, either ensure a successful investigation or doom it to failure. Learning the proper steps for organizing a failure investigation ensures success.

Failure investigation crosses company functional boundaries and is an integral component of any design or manufacturing business operation. However, a poorly organized investigation may not provide the necessary information to solve the manufacturing problem or assist in a redesign.

The Failure Analysis Process

Failure analysis is a process for determining the causes or factors that have led to an undesired loss of functionality. Most failure analyses primarily address failures of components, assemblies, or structures, and the approach is consistent with the knowledge base of a person trained in materials engineering.

Over the past few decades, materials engineers have greatly helped to advance the scientific foundation of failure analysis. Many people still define the causes of failure in a rather binary manner: Was the part defective or was it abused? Obviously, there are many types of defects, including those due to deficient design, poor material, or manufacturing mistakes. Whether such "defects" exist in a given component that is undergoing failure analysis often can be determined only by someone with a materials background—because many defects are visible only with the aid of a microscope. While microscopes may be widely available, the knowledge and experience required to interpret the images is not. Increasingly powerful scanning electron microscopes have helped provide a more fact-based foundation for opinions that may have been heavily speculative in the past.

Design-related defects may require assessment by a materials engineer, as many design engineers are not very familiar with material factors such as the natural variations within a material grade. Every "failed" object is made of some material, and some common materials can lose more than 90% of their usual strength if not processed properly. Clearly, prior to reaching a conclusion as to the most significant causes of the failure, someone must determine whether the correct material was used and whether it was processed properly. This often requires both an investigation of documentation and a series of physical tests.

Underlying Causes of Failure. Through work on spectacular failures and on failures that have caused great pain and loss, materials engineers have been led to ask deeper and broader questions about the underlying causes that lead to failures. In many cases it becomes clear that there is no single cause or no single train of events. Instead, factors combine at a particular time and place to allow a failure to occur. Sometimes the absence of one factor may have been enough to prevent the failure. Sometimes, however, it is impossible to determine, at least within the resources allotted for the analysis, whether any single factor was key.

Failure analysts must look beyond the simplistic list of causes that some people still promote. They must keep an open mind and always be willing to get help from other experts. Many beginning failure analysis practitioners may have their projects defined for them when they are handed a small component to evaluate, and thus may be able to follow an established procedure for the evaluation. This is especially true when working for an original equipment manufacturer where much prior experience and knowledge of the physical factors that tend to go wrong with a component have led to established procedures. In that case, a particular analysis may not require extensive pretesting work. However, for the practitioner who works in an independent laboratory or must look at a wide variety of components, following a predefined set of failure analysis instructions may prove inadequate. Established "recipe"-type procedures are generally inadequate for the more advanced and broad-minded practitioner as well. A broad, systematic methodology is more appropriate.

Reasons behind the failure of a component, assembly, or structure can be multilevel. In other words, a failure should not be viewed as a single event. The actual physical failure—a fracture, an explosion, damage by heat or corrosion—is the most obvious. However, other levels of failures generally exist that allow the physical event to happen.

Consider the case of a simple failure whose direct physical cause was an improper hardness value. However, one or more persons allowed the improperly hardened component to be manufactured and used. Human factors generally are very difficult to investigate within a manufacturing organization, because cultures that allow a particular type of failure to occur usually will not have systems in place that enable simple remedies to be enacted for deeper-level causes. For example, if someone in an organization wants to investigate causes beyond the simple fact of improper hardness, it may be discovered that the inspection clerk was not properly trained to note reported hardness values when receiving materials. Changing a corporate culture to include better training and education is often difficult; many corporations are structured so that the people who are responsible for training do not have an open line of communication with those doing the investigation. This only increases the difficulty of implementing change to prevent failures.

Failure Prevention. Failure analysis of a physical object is often only part of a larger investigation intended to prevent recurrences. When taking the broadest view of what is required to prevent failures, one answer stands out: education. To reduce the frequency of physical failures, education must be instilled at multiple levels and on multiple subjects within an organization.

Education, of which job training is a single component, is what allows people at all levels of an organization to make better decisions in time frames stretching from momentary to career-long. Many books contain exercises that help the reader to restructure knowledge into a more useful and accessible form. Other books help the reader learn to recognize incorrect lines of reasoning; an excellent example is *Tools of Critical Thinking: Metathoughts for Psychology* by D. Levy (Ref 1).

Specific levels of failure causes (Fig. 1) have been defined by Failsafe Network (Ref 2) as:

- *Root:* the true cause of failure, encompasses the next three items
- *Physical:* the failure mechanism (fatigue, overload, corrosion, etc.)
- *Human:* the human factors that lead to the physical cause
- *Latent:* the cultural/organizational rules that lead to the human cause

Clearly, many people involved with failure analysis incorrectly use the term "root cause" when what they really are referring to is a simple physical cause.

If failure analysis tasks are performed adequately, then the analyst ultimately should be able to list the causes found, show that the failure would have happened the way it did, and also show that if something different

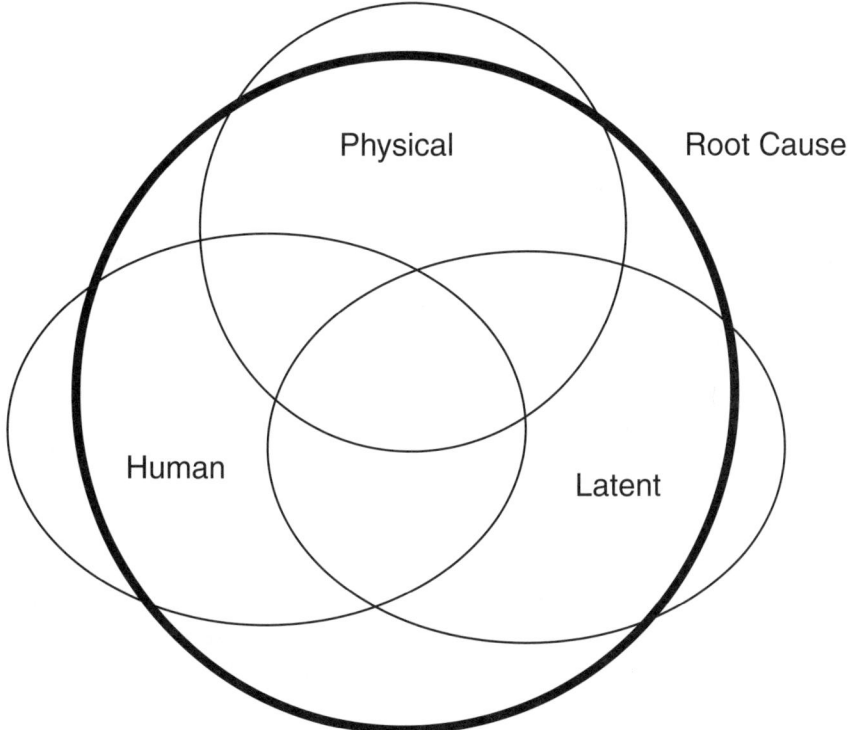

Fig. 1 Levels of failure causes

had happened at some step along the way, the failure would not have occurred or would have occurred differently. Unfortunately, this definitive demonstration of the failure is not always possible. Even a lengthy and thorough investigation can result in unknowns. The honest analyst is left to make a statement of the factors involved in allowing conditions that promoted the likelihood of failure. This is still a useful task, perhaps more useful than one that merely pins "blame" on a particular individual or group. Understanding the factors that promoted a failure can lead to an understanding of exactly what is required to improve the durability of products, equipment, or structures. Understanding goes beyond knowledge of facts. Understanding requires integration of facts into the knowledge base of an individual so that the facts can be transformed into product and/or process improvement.

By now it should be clear that failure analysis is a task that requires input from people with many areas of expertise. A simple physical failure of a small object may be analyzed by a single individual with basic training in visual evaluation of engineered objects. However, going to the level of using the failure analysis to improve products and processes requires expertise in the various aspects of human relations and education, at the least. Failure analysis of a complex or catastrophic failure requires much more.

Legal Ramifications. People who perform failure analysis as part of their job function need to be aware of how their legal obligations are defined. Investigators who perform destructive testing on a failed component may sometimes be held personally accountable for the destruction of evidence. Company employees must learn to protect themselves. Investigators who were "just doing their job" have been successfully sued by parties that the judicial system determined had a legitimate interest in the outcome of the failure analysis project.

Never unquestioningly agree to destructively test a "failed" component. However, this places the destructive testing technician or engineer in a troublesome position, as it is sometimes difficult to see that a component has failed. Even if that information is given, relevant background details are often difficult to obtain, especially in a highly structured and hierarchical corporate culture. Pressure to finish the analysis quickly is common. People who request failure analysis work may not be aware that rushing ahead into the destructive portion of an investigation may well destroy much information and evidence.

Those who perform failure analysis work must realize that many people are still unaware of what failure analysts have to offer in terms of allowing clients or fellow employees to replace speculation with facts. This book is intended to demonstrate proper approaches to failure analysis work. The goal of the proper approach is to allow the most useful and relevant information to be obtained. All the valid approaches require planning, defining of objectives, and organization prior to any destructive testing.

Simultaneous preservation of evidence also is required. It should now be clear that proper failure analysis cannot be done with input from only a single individual. Even someone only participating in the "straightforward" portions of the investigation of physical failure needs to know how his or her contribution fits into a bigger picture.

The competent failure analyst needs to know more than the failure analysis process and the tools used to support it. The competent failure analyst needs to understand the function of the object being analyzed and to be familiar with the characteristics of the materials and processes used to fabricate it. The failure analyst needs to understand how the product was used and the culture in which it was used. Communication skills are a must. When you ask a question, do you know for certain what the answer "yes" means? In some cultures, the word "yes" means "I heard the question" and does not imply that the answer is actually affirmative. The failure analyst must always be well versed in multiple disciplines.

Recognizing a Failure

A good basic definition of a failure is the inability of a component, machine, or process to function properly. However, this definition is not limited to things that break, causing a complete shutdown of a system. The concept of a failure is much larger. Can you identify a failure if you see one?

Does the situation in Fig. 2 appear to be a failure? Would you put your child in the sink for a bath with dishes? On the other hand, the child *does* appear to be on one side and not actually in with the dirty dishes.

Fig. 2 Is this a failure?

Every one of us has watched a sporting event from the Olympics to Little League in which we see athletes succeed and fail in their endeavors. The sight of a dejected athlete looking at the ground with head in hands is a universal sign of failure. No one has to tell us the athlete failed, we know it.

The beaver in Fig. 3 might be a victim of foul play—or, perhaps, a victim of overconfidence. How often has a supplier or coworker told you, "I know what I'm doing, you don't have to warn me." Next thing you know, there is a hole in a part where it doesn't belong.

I could never determine if the photograph in Fig. 4 was an advertisement for the strength and durability of Mercedes-Benz trucks or a poster for traffic safety. I will always wonder where these folks were, how far they got, and if the man at the lower left got a lift or not. The trucks are obviously overloaded; it is only a matter of time before they will fail. Of course, no sensible person would run a machine over its capacity, right? Would a manufacturer run its machines beyond capacity just to get a few more parts per second?

In many failure investigations, the failure is represented by a broken piece of hardware like the shaft in Fig. 5. Anyone can look at this shaft and determine that it failed. In addition, the effect to the equipment is an instantaneous shutdown. However, that is not the only way by which failures occur.

Fig. 3 Even an expert can sometimes fail.

Fig. 4 A failure waiting to happen

Consider the propeller in Fig. 6. How did it fail? At first glance, some might assume that the propeller operates in salt water and thus failed due to corrosion of some kind. But look closely. The damage occurred only on the leading edges of the blades. The true failure mechanism is cavitation. You might think to yourself, "How was I supposed to know that? I'm not an expert on propellers." To perform a failure investigation, you must become knowledgeable about the component or structure and all possible failure modes. And you need to consult with experts in the field.

As noted previously, a good definition of a failure is the inability of a component, machine, or process to function properly. The examples presented thus far indicate that a failure investigation may involve factors

Fig. 5 Failed shaft

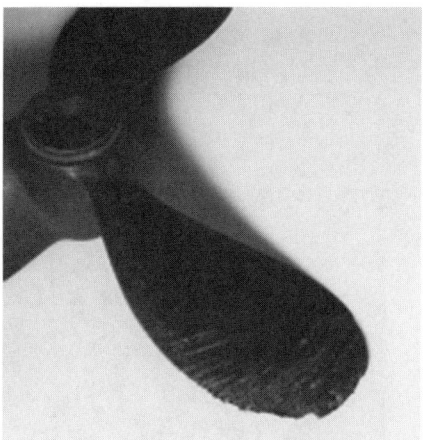

Fig. 6 Propeller that failed by cavitation

other than a broken component. Let us look at other, more detailed examples.

F-20 Tigershark. In the early 1980s, Northrop developed the F-20 Tigershark (Fig. 7) in response to a U.S. government interest in the private development of a tactical jet fighter specifically tailored to meet the security needs of allied and friendly nations. Unfortunately, the F-20 Tigershark fighter was produced without a contract from the military.

The F-20 was inexpensive, reliable, and easy to maintain. Based on comparisons with contemporary international fighters, the F-20 consumed 53% less fuel, required 52% less maintenance labor, incurred 63% lower operating and maintenance costs, and offered four times the reliability. In an era when comparable aircraft cost $15 to $30 million apiece, Northrop designed one that cost about $1 million by eliminating all the technical bells and whistles of other planes. It must have sold like hotcakes, right?

The company never sold a single F-20. Why? None of the allied and friendly nations had asked for it. The U.S. military did not want it, either. The military was used to the concept that it told the aircraft industry what it wanted, and companies bid on the project. No one was used to the aircraft industry simply providing an aircraft. The F-20 offered fuel efficiency, cost efficiency, and better reliability, but nobody had asked for those characteristics.

The last time I read about the F-20, only five still existed. To make matters worse, one of the primary reasons the company was not able to sell the aircraft to overseas allies was that the United States armed forces had refused to buy the aircraft. The U.S. military purchased the F-16 instead, and the overseas allies decided to buy the same aircraft. It could be argued that the F-20 was a marketing failure.

The International Space Station (ISS) construction was initiated on November 20, 1998 with the launch of the *Zarya* control module, and phases I and II of its three-phase construction were completed on July 15, 2001 with the addition of the *Quest* airlock. Permanent human habitation has been in progress since the arrival of the Expedition 1 crew on October 31, 2000. The Expedition crews orbit the earth every 90 min at an altitude of 354 km (220 miles) while traveling at 7,823 m/s (17,500 mph). As of

Fig. 7 F-20 Tigershark, circa 1982

January 26, 2005, the resident of the ISS is the Expedition 10 crew, who arrived in October 2004. Between December 7, 1998 and January 26, 2005, there have been 57 extravehicular activities (EVAs), or "space walks," totaling 343 hours, 45 minutes (Ref 3), in the assembly of the ISS (Fig. 8). In comparison, since the first space walk by astronaut Ed White in 1964 until 1998, NASA astronauts had completed a total of 377 hours of space walks.

Unfortunately, astronauts lose items during EVAs. Interestingly, NASA does not consider this a failure unless the astronaut fails to complete the designated task. Astronauts have lost entire tool bags filled with costly beryllium tools specially designed for lightweight and stiffness. The bags simply floated away.

The astronauts are supposed to operate on a dual-anchor system for themselves and their equipment. When they move, they undo one anchor, move, set the anchor, then go back and move the second anchor. However, as the EVA progresses over a period of hours, the astronauts begin to

Fig. 8 Astronauts perform an EVA at the space station

move with a little more confidence. One reported video shows an astronaut diligently working on his task, while his bag slowly floats up and away. He turns, grabs it, and brings it back. He keeps working, and the bag starts to float away a second time. Once again he grabs it and brings it back. The third time, the bag floats out of reach. I guess two out of three is not too bad, but the problem would have been avoided had he just anchored his tool bag.

Tool bags are not the only items lost. While performing tasks, astronauts stand on 1.2 by 1.2 m (4 by 4 ft) aluminum reaction pads. Believe it or not, they have lost the pads and now carry extra ones on each mission. Is that a failure? Once again, NASA does not see it as such unless the astronaut cannot complete the task.

In the larger picture, the loss of tool bags and reaction pads is a failure because these items are now space debris. Like the ISS and the Space Shuttle, the space debris also travels at 7,823 m/s (17,500 mph)—though not necessarily in the same direction or at the same altitude. We can currently track more than 9000 pieces of such debris, and we just keep adding to it.

The ramification of this space debris is that if the Space Shuttle or the ISS was to be hit by a 1.2 by 1.2 m (4 by 4 ft) piece of aluminum, significant damage would result. Years ago, a paint chip the size of a postage stamp hit the Space Shuttle *Columbia,* cracking the windshield. So is a lost tool bag or reaction pad a failure? NASA says no, but I believe it is.

California Energy Crisis. The energy crisis in California that started at the beginning of the millennium is a political and economic failure of enormous magnitude. The crisis originated from state government legislation to deregulate electric power utilities and to open up the market to competition. However, a series of subsequent events led to an energy crisis that was unprecedented in the history of the state. Large energy companies went bankrupt, the cost of utilities skyrocketed, and many politicians, including the governor, lost their jobs. Energy bills increased as much as 40 to 60% within a short period of time. In 2000 and 2001, rolling blackouts affected thousands of electric customers in various areas of the state, all because the state government decided to deregulate and did not really understand the potential consequences of its actions.

The decisions that the state government made were not illegal or even necessarily wrong, but the politicians just did not think through all likely outcomes. They saw other approaches to deregulation work in other states and thought it was a great idea. But some companies, such as Enron, found a way to take advantage of the situation. The ramifications of this failure were widespread. The result was a catastrophic energy crisis. The governor, Gray Davis, was ousted in a recall election, and the state accumulated tens of billions of dollars in debt, with predictions of debt reaching $80 billion by 2010.

How did this happen? Where was the failure? When you read about it in the newspaper or on the Internet it seems very simple. The original bill easily made its way through the legislature and attracted little public attention. The governor and the legislators thought it was a good idea. However, the outcome is that the state of California will be suffering for the next several decades. The California energy crisis can be considered a failure of a state government, or "machine," to control its energy sources. Or, is it a failure of the citizens to control their state government?

USS *Greeneville*. In February 2001, the USS *Greeneville* SSN 772 (Fig. 9) surfaced in the middle of the ocean, far from any land mass. A Los Angeles class nuclear attack submarine launched in February 1996, it was one of the U.S. Navy's most sophisticated and technically advanced vessels. We live on a planet that is three-quarters ocean, with vast areas of open space. So how does a brand-new, 110 m (360 ft) nuclear submarine surface and ram a 55 m (180 ft) Japanese fishing vessel called the *Ehime Maru*?

Some people wanted to blame the navy's practice of allowing dignitaries aboard certain ships. Sixteen civilian visitors were on board the USS *Greeneville*. The theory was that a civilian was distracting the sailor steering the submarine, or perhaps the civilian was even operating the submarine. In this case, the failure was blamed on navy policy. Another theory was to blame the sonar operator. How could the sonar operator not see a fishing trawler in the middle of the ocean? A third theory was that the

Fig. 9 USS *Greeneville*

Ehime Maru was not really a fishing vessel but a high-tech submarine chaser.

So what was the root cause of the failure? Was it a failure of navy policy? Was it a failure of naval training? Was it a failure of sonar operation? Was it a navigational miscalculation? The truth may never be known, but the ramifications are clear. Almost all the personnel on the fishing vessel were killed. The United States spent millions of dollars to retrieve their bodies from the bottom of the ocean, and the country's prestige suffered a blow. The 42-year-old submarine commander took full blame for the incident, thereby ending his career as commander of a ship.

It turns out that ramming incidents such as this are not uncommon in the navy. Naval ships have rammed into each other in the middle of the ocean. Ships making turns in harbors have severely damaged their bows. In fact, the USS *Greeneville* was involved in a similar incident in January 2002 when it collided with the USS *Ogden* transport vessel. The two vessels were attempting to get into position to transfer two sailors going home on emergency leave. Is this a failure in the way the navy trains its officers?

So where is the failure? Who was responsible? Was it a failure of training of the commander and crew? Was it a failure of operation of the submarine sonar? Was it a miscalculation of the navigation officer? Was it a failure caused by the policy of the navy to allow visitors on vessels, which can add confusion and distraction?

The welded crankshaft shown in Fig. 10 failed in an automobile. The mechanic had probably told the customer, "You know, I can weld that crankshaft up good as new. No sense replacing it just because of a small crack. I can save you some money." That seemed like a good idea until the crankshaft came apart at 18 m/s (40 mph). What began as a small noise in the engine turned into a completely destroyed engine. The cus-

Fig. 10 Welded crankshaft that failed

tomer did not know anything about engines or welding, but he liked the idea of saving money. The mechanic apparently did not know anything about welding or metallurgy either.

The gears shown in Fig. 11 are part of the gearbox from a large piece of land-moving equipment. The company uses hundreds of these parts for new and rebuilt gearboxes. One day a cost analyst said, "We can save a ton of money if we replace the forged gears with these machined ones." He had been contacted by a new supplier who offered a great price. He did not know the technical aspects of the gears or why the forging process was important, but he liked the idea of saving money. It was a great idea until the gears began failing at 30% of the expected life and the company began spending a lot more money rebuilding gearboxes under warranty.

Titanium Domes. This example shows how scheduling issues—and the tendency for people to cut corners to save time—can contribute to failures.

Titanium domes like those in Fig. 12 are frequently reduced in size by a process called chemical milling. The domes are actually a very simple

Fig. 11 Failed gears

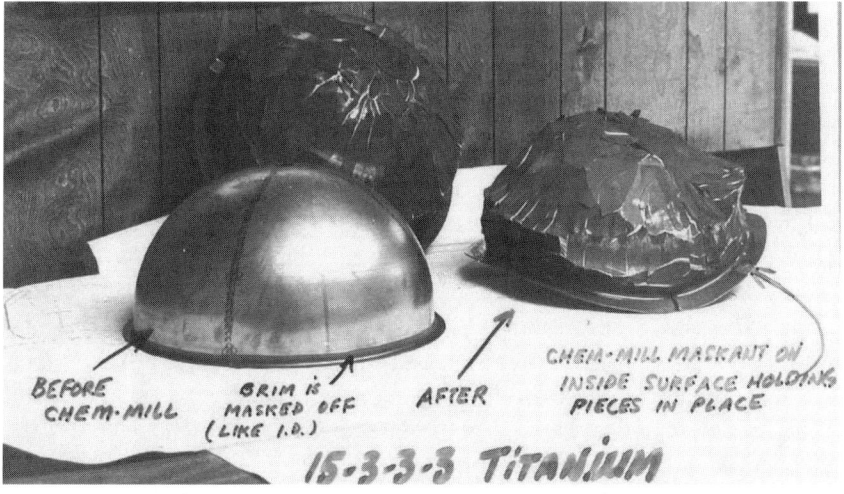

Fig. 12 Titanium domes. Before chemical milling (left) and after chemical milling (right) in an incorrect etchant that caused the dome to undergo brittle fracture

forging produced by back extrusion. End-users machine the forgings to a certain size and then chemical mill them as a secondary operation. The domes end up with less residual stress from the chemical milling process than from the machining process.

Chemical milling basically involves placing the titanium domes into a bath of acid, which eats away the titanium. In more correct terms, this process uses a chemical etchant to reduce the thickness of the material. However, the proper etchant must be used or the material will be affected. In addition, a maskant frequently is used to prevent material removal from certain areas.

One day a chemical milling supplier was visited by a customer with three titanium domes. The customer explained that he had to have the parts chemical milled immediately. The domes were made from titanium 15-3-3-3 (Ti-15V-3Al-3Cr-3Sn), a special alloy designed for spin forming. The supplier told the customer that he could not do chemical milling of the domes because he currently had the wrong etchant in the tank. The chemical milling process drives hydrogen into titanium, and there are limits to the amount of hydrogen each titanium alloy can absorb. Titanium will become very brittle if the chemical milling process is not properly controlled.

The customer insisted that the job be performed immediately, regardless of the etchant. So the supplier masked the outside surface of the first dome and placed it in the chemical milling bath. After a few minutes, a popping noise was heard. The dome had cracked into so many pieces that the only thing holding it together was the maskant. The titanium had absorbed too much hydrogen from the bath and had become so brittle that it was breaking under the residual stresses from the heat treating and machining processes. Even though the first dome was destroyed, the customer insisted that the second dome be placed in the chemical milling bath in order to save time. It, too, cracked. The customer took his last dome back home and returned later that week when the chemical milling supplier had the correct acid in the bath. The physical cause of failure was use of the incorrect etchant, but the root cause was unrealistic scheduling demands.

Tire Failures Leading to Rollovers of Sport Utility Vehicles. The failures of Firestone tires on sport utility vehicles (SUVs) such as the Ford Explorer have caused accidents that have cost many lives. These accidents have occurred over a period of several years. In May 2000, the U.S. National Highway Transportation Safety Administration formally requested information from Ford Motor Company and Bridgestone Firestone about the high incidence of tire failure on Ford Explorer vehicles. Later that year, Ford obtained and analyzed the data on tire failure. The data revealed that 15 in. ATX and ATX II models and Wilderness AT tires had very high failure rates. The failures involved tread peeling off. Underinflation of the tires appears to have been a contributing factor. In many cases, tire failure at high speeds led to the vehicle rolling over, resulting in injury or death to the occupants. Ford blamed the tires and Firestone blamed the SUVs.

In June 2001, *USA Today* reported that Firestone was planning to replace 13 million tires. At approximately $100 per tire, Firestone was facing a loss of 1.3 billion dollars. As of the summer of 2002, it was reported that Firestone had replaced 10 million tires at an estimated cost of one billion dollars. No one has ever conclusively determined who is at fault. Ford continues to sell many Explorer SUVs.

Let us look at another interesting angle on this situation. The replacement tires Firestone was providing, or selling, were the Wrangler HT and the Grabber APXL. Firestone data had indicated that the Wrangler HT had a failure rate of 13.7 per million tires and the Grabber APXL had a failure rate of 10.9 per million tires. The Firestone maximum failure rate criterion is five incidents per one million tires. These two replacement tires did not meet the internal Firestone requirements. What kind of failure is this situation? How is this happening? Firestone is selling two tires that do not appear to be any better than the original. This is certainly not a failure of marketing.

This failure has many potential root causes and complications, including the quality of the tires and the design of the SUV. Another factor to consider is the driving habits of the SUV owners. Is this one of the reasons for the problems? People have a tendency to drive SUVs like they are sports cars. Many SUVs have engines that enable them to accelerate like a sports car, but not the stability, handling, and braking capability of a sports car. It is much harder to drive and control a large, trucklike SUV, especially with underinflated tires, than a car. In addition, SUVs have replaced minivans and station wagons as the family vehicle of choice for many Americans. The chance that a driver is not paying adequate attention to the road increases with the number of occupants in the vehicle, especially if some of the passengers are small children. The combination of overpowered vehicles with insufficient stability, handling ability, and braking systems being operated by people who may not be paying full attention should also be considered.

This is a very complicated issue. It is very difficult, if not impossible, for all parties to determine who is to blame, and then to agree on decisive corrective actions. The popularity of SUVs with consumers—and their profitability for automobile manufacturers—also should be considered. Manufacturers of SUVs make profits on the vehicles in two ways. One, the popularity of SUVs makes them high-selling models. Two, the classification of SUVs as trucks for the purposes of EPA fuel efficiency standards allows manufacturers to sell more vehicles without implementing potentially expensive fuel efficiency technologies. Over the last few years, there has been a movement to put SUVs into the car category for EPA ratings. So far, automobile manufacturers have successfully fought this change.

So is the real issue that the Firestone Tire-SUV rollover problem is an *unpopular* failure? Is the real failure that no one wants to tell the American public that their favorite SUVs are not safe and the way they are driven is not safe? Or that no one wants to tell the powerful SUV and tire manufacturers that their products are not as safe as they should be?

Public opinion may be changing. In December 2003, newspapers reported that the automobile manufacturers agreed to cut SUV injury risks (Ref 4). Studies had determined that when SUVs rollover their roofs collapse, indicating that they are not very structurally safe. Different public watchdog groups were set to produce commercials saying the SUVs are not safe and that the government should set standards for rollover stability. The direction appears to be not to fix the problem that causes SUVs to rollover, but to make them safer when they do rollover.

This kind of coordinated bad publicity forced the automobile manufacturers to react. Their response was to volunteer to increase the safety of SUVs in rollovers. Why are they doing it voluntarily? Because this approach will allow them to set the new safety criteria. If the government makes a law that says SUVs have to meet certain criteria, the automobile companies have to meet those criteria. If the cost to meet those criteria goes through the roof, the manufacturers lose their profits. But if the automobile companies set the new criteria, they can make sure it is done so that they can do it profitably. This approach does nothing to fix the root cause of the failure, but will allow more people to live through the failure if it occurs. It is treating the symptom, not the disease.

Space Shuttle Disaster. On February 1, 2003, the Space Shuttle *Columbia* exploded during reentry after a scientific mission. The Space Shuttle fleet is considered to be a national treasure, and the loss of the *Columbia* most likely will affect the United States space program for the next 25 years. What was the root cause of the failure? Was it the part—the wing or the piece of foam from the external tank that struck the wing? Was it the machine—the inability of the *Columbia* to sustain a debris strike during launch or to reenter the Earth's atmosphere with the damage from the strike? Was it the assembly of the shuttle by NASA engineers? Was it the inability of these engineers to examine and analyze the damage? Or was it the NASA decision-making process—the inability of NASA to decide on the best course of action within the time frame of the shuttle mission?

The Columbia Accident Investigation Board (CAIB) spent six months and millions of dollars identifying the root cause of the failure. The physical root cause was determined to be a breach in the thermal protection system on the leading edge of the left wing. The breach was initiated by a piece of insulating foam that separated from the left bipod ramp of the external tank and struck the wing in the vicinity of the lower half of a reinforced carbon-carbon panel at 81.9 s after launch. Twelve days later, during reentry, superheated air penetrated the leading-edge insulation and progressively melted the aluminum structure of the left wing, weakening the structure until increasing aerodynamic forces caused loss of control, failure of the wing, and breakup of the Orbiter.

However, this constituted only the physical root cause of the accident. In the end, the CAIB issued various findings and 29 separate recommendations (Ref 5). Here is the first part of the official release on August 26, 2003:

> The CAIB report concludes that while NASA's present Space Shuttle is not inherently unsafe, a number of mechanical fixes are required to make the Shuttle safer in the short term. The report also concludes that NASA's management system is unsafe to manage the shuttle system beyond the short term and that the agency does not have a strong safety culture.
>
> The Board determined that physical and organizational causes played an equal role in the *Columbia* accident—that the NASA organizational culture had as much to do with the accident as the foam that struck the Orbiter on ascent. The report also notes other significant factors and observations that may help prevent the next accident.
>
> The Board crafted the report to serve as a framework for a national debate about the future of human space flight, but suggests that it is in the nation's interest to replace the Shuttle as soon as possible as the primary means for transporting humans to and from Earth orbit.
>
> The Board makes 29 recommendations in the 248-page final report, including 15 return-to-flight recommendations that should be implemented before the Shuttle Program returns to flight.

As an example, here is Recommendation 4, released July 1, 2003:

> Upgrade the imaging system to be capable of providing a minimum of three useful views of the Space Shuttle from liftoff to at least Solid Rocket Booster separation, along any expected ascent azimuth. The readiness of these assets should be included in the Launch Commit Criteria for future launches.
>
> Consideration should be given to using mobile assets (ships or aircraft) to provide additional views of the vehicle during ascent.

The CAIB felt there was not enough information provided during the launch for engineers to properly assess the damage to the Shuttle wing. Decisions could have been made at the beginning of the mission when the Shuttle had more fuel and time. It seems like such a small issue in comparison to the tragedy that occurred almost two weeks later. However, the smallest of root causes can initiate a cascade of events that lead to tragedy.

Summary. Hopefully, this chapter has opened your eyes to the fact that there are all kinds of failures and the first step is to recognize a failure when you see one.

ACKNOWLEDGMENT

Portions of the section "The Failure Analysis Process" in this chapter have been adapted from Debbie Aliya, The Failure Analysis Process: An Overview, *Failure Analysis and Prevention,* Volume 11, *ASM Handbook,* ASM International, 2002, p 315–323

REFERENCES

1. D.A. Levy, *Tools of Critical Thinking: Metathoughts for Psychology,* Allyn and Bacon, 1997
2. C.R. Nelms, Failsafe Network, Inc., Montebello, Virginia
3. NASA Human Space Flight Web site, National Aeronautics and Space Administration, spaceflight.nasa.gov/station/isstodate.html (Web site accessed Feb 2005)
4. *The Plain Dealer,* Cleveland, Ohio, Dec 4, 2003
5. Columbia Accident Investigation Board, www.caib.us (Web site accessed Feb 2005)

Case History: BOAC Comet Crashes, Mediterranean Islands of Elba and Stromboli (1954)

Robert J. Serling

Two new jetliners exploded in midair; the cause was eventually traced to metal fatigue cracks around a small window in the upper fuselage that developed during pressurization and depressurization cycles.

Background

At 10:50 on the morning of January 10, 1954, a British Overseas Airways Corporation (BOAC) de Havilland Comet en route from Rome to London radioed Rome that it was breaking through the overcast at 8,230 m (27,000 ft) and climbing to an assigned cruising altitude of 10,973 m (36,000 ft). Less than two minutes later, fishermen off the islands of Elba in the Mediterranean saw flaming wreckage falling out of the clouds and into the sea.

The last word from the ill-fated aircraft was a routine message to another BOAC flight, reporting on the height of the cloud cover. Whatever destroyed G-ALYP (the aircraft's registration letters, translated into "George Yoke Peter" as its air traffic control letters) interrupted the report in mid sentence and happened with devastating swiftness. There were 29 passengers and six crew members aboard.

The jet age had been born two years earlier with the takeoff of the same aircraft on a flight to South Africa, a flight that had also inaugurated BOAC's scheduled jet service. The original Comet was only a 36-passenger aircraft (40 on longer-ranged version with extra fuel tanks), but its speed and ability to fly above virtually all weather revolutionized air travel. Britain was understandably proud of this technological triumph, for in 1952 Russia was two years away from launching its TU-104 (actually a converted bomber), and in America, the Boeing 707 and the Douglas DC-8 were still on the drawing board. At the time of the mysterious Elba tragedy, de Havilland had firm airline orders for 50 Comets, and negotiations for another 100 were well underway.

During the first two years of the enormously popular Comet airline service, there had been two takeoff accidents, one nonfatal but the other resulting in the death of everyone on board. Each was attributed to the pilot's attempt to rotate (lift) the nose before adequate speed had been attained. In a third accident, a BOAC Comet flew into a violent thunderstorm and apparently disintegrated as a result of lightning and turbulence related to the storm. But neither fatal crash had caused any alarm bells to ring; pilot error and storm-caused structural failure were nothing new to aviation. Flaws in the Comet's design were not considered factors in these accidents.

In fact, cognizant that the new jetliner would be operating in an environment of unprecedented speeds and altitudes, chief designer Ronald Bishop and his engineering staff were particularly aware of pressurization stresses on an aircraft structure. Cabin pressurization—a means of compressing thin upper-atmosphere air into heavier, more breathable air as it enters the cabin—dated back to the Boeing Stratoliner of 1938. But there was a vast difference between a piston-engine plane's pressurization/depressurization cycles and those of a jet, which not only climbed and flew twice as fast, but routinely cruised 3,048 m (10,000 ft) higher.

On transport aircraft like the Constellation and the DC-6, fuselages—the central compartments of airplanes that house the crew, passengers, and cargo—were designed to safely hold maximum pressure of about 0.028 MPa (4 psi), sufficient to bring an aircraft's actual altitude of 7,620 m (25,000 ft) down to a level of air pressure in the cabin equivalent to that of an altitude of less than 1524 m (5000 ft). The Comet required 0.055 MPa (8 psi) to achieve the same breathable air, but British civil aviation authorities insisted that the new jet's cabin structure cope with more than 0.110 MPa (16 psi). Bishop and his chief structural engineer, Robert Harper, went beyond even that safety margin; the Comet fuselage was designed to hold 0.138 MPa (20 psi).

To avoid excessive weight, the engineers chose a relatively thin gauge for the aluminum skin. It was only 0.711 mm (0.028 in.) thick, not much thicker than a postcard. Another critical and controversial decision involved the windows: the frames were square, the durability of which the U.S. Civil Aeronautics Administration (CAA) questioned. This predecessor agency of the Federal Aviation Administration (FAA), charged with the responsibility of approving the airworthiness of American-operated Comets, suggested that oval-shaped frames would distribute pressurization stresses more equally. Engineers from BOAC responded that the Comet's windows had been tested at up to 0.70 MPa (100 psi) of pressure without any signs of fatigue.

Following the Elba disaster, BOAC voluntarily grounded its entire seven-aircraft Comet fleet for inspection as a precautionary measure while investigators pondered all possibilities. The causal candidates included sabotage; an explosion caused by a ruptured turbine blade penetrating a fuel tank; structural failure from clear air turbulence; an in-flight engine fire that either ignited fuel or weakened the structure to the point of failure; explosion of hydrogen from a leaking battery; or an explosion of fuel vapor in an empty tank.

One BOAC technical expert wondered if metal fatigue might have caused explosive decompression. This seemed impossible to other BOAC officials. The aircraft had logged fewer than 4,000 hours of flying time, the equivalent in mileage of an automobile only a few months off the showroom floor, and the de Havilland designers had assured BOAC that the Comets' structures could endure the stress of 10,000 flights' worth of repeated pressurization and depressurization.

What made it so difficult to determine the cause of the crash was the lack of solid clues. Royal Navy salvage crews were groping for wreckage buried 152 m (500 ft) deep and scattered over 259 km^2 (100 square mi.) of Mediterranean sea floor. Meanwhile, a special investigating committee recommended some 50 modifications to the Comet, all preventative measures based on possible but unproven midair explosion theories. Fuel lines were strengthened. Amor-plated shields were installed between the engines and fuel tanks. Improved smoke and fire detectors replaced original equipment. By late March, BOAC's Comet fleet was back in the air even as the salvage crews continued their search.

These measures were taken in vain, however, for on the night of April 8, 1954, a second Comet aircraft explosion occurred, resulting in 21 deaths. The aircraft G-ALYY ("George Yoke Yoke"), flying from Rome to Cairo, was climbing to its assigned altitude of 10,820 m (35,500 ft) when radio contact suddenly ended. The following morning, five bodies and two aircraft seats were recovered from the sea near the island of Stromboli, and for the second time in less than four months all Comets were grounded. Prime minister Winston Churchill ordered Britain's most prestigious aeronautical science organization, the Royal Aircraft Establishment (RAE), to take over the investigation of both accidents. The RAE investigation, headed by Sir Arnold Hall, was responsible for attributing the cause for the destruction of the world's first jetliner and, in effect, possibly determining the fate of all future jet travel.

Details of the Disaster

The RAE scientists were aided immensely by the salvage miracle performed by the Royal Navy. There was little hope of recovering Yoke Yoke's wreckage, which had fallen into water 1067 m (3500 ft) deep. But less than a month after Yoke Yoke's disappearance, the navy had recovered about two thirds of Yoke Peter's wreckage. The remains started to arrive at Farnborough, the RAE's main research facility, via ship, at too slow a pace for Arnold Hall. Without waiting for official permission, he commandeered a huge U.S. Air Force cargo plane to speed up wreckage delivery and inspection of every salvaged scrap.

Almost from the start, Hall suspected metal fatigue and so did Dr. Peter Walker, head of RAE's Structural Department. The investigation detoured temporarily down one blind alley after the discovery that one of Yoke Peter's engine turbine blades was missing, a finding that lent credence to the theory that a severed blade had punctured a fuel tank. But this possibility was discarded after the turbine casing was found intact and it was determined that the blade had been torn off by crash impact forces.

The final results of the autopsies performed in Italy on the 15 bodies recovered from Yoke Peter and the five found near the site of the Yoke Yoke crash suggested to the RAE that all had been victims of "violent movement and explosive decompression." The Italian pathologist further concluded that the traumatic ruptures of the hearts and lungs had occurred before impact into the sea. This report supported the RAE investigators' suspicions that the pressure cabin had failed prior to the explosion.

Hall ordered the construction of a tank large enough to hold a Comet fuselage. One of the grounded jetliners was placed inside with its wings protruding from a hole on each side of the tank. Hall believed the original pressurization tests on the prototype had not been sufficient. Although they had involved repeated pressurization and depressurization cycles to determine the effect on the aircraft's structural life, the tests had not been accompanied by the use of hydraulic jacks to simulate the motion of an aircraft in flight.

The RAE installed such jacks to flex the wings of the submerged Comet in conjunction with the hydrostatic tests. The fuselage was filled with water (instead of air because water would cushion an explosive decompression), and the pressure was raised to slightly over 0.055 MPa (8 psi)— the Comet's normal pressurization at 1067 m (3500 ft). The level of 0.055 MPa (8 psi) of pressure was maintained for three minutes in unison with the up-and-down motions of the wings, then reduced to 0 MPa (0 psi), and then increased back up to 0.055 MPa (8 psi) for another three minutes.

Each cycle, corresponding to the pressure variations of a three-hour flight, was repeated around the clock until the airframe had accumulated the equivalent of 9000 flying hours or 3000 flights. The hydrostatic test schedule had aged the airframe 40 times faster than would have been possible in normal airline service.

In late June, just as the 9000-hour mark was reached, the fuselage could no longer maintain the pressure and a crack developed. After the water was drained from the tank, a crack 2.4 m (8 ft) long could be seen along the top of the fuselage, extending from a fracture in one corner of a small escape hatch window and bisecting a cabin window frame. Metallurgic inspection of the hatch window revealed the discoloration and crystallization associated with metal fatigue.

This evidence was corroborated when in mid-August a section of Yoke Peter's fuselage roof was recovered and flown to Farnborough. One look provided confirmation of the hydrostatic tests. In the corner of a navigation window atop the fuselage was a telltale crack that had grown into a wide split. With the appearance of this final clue, the RAE investigators eventually came to the conclusion that the accidents were primarily "caused by structural failure of the pressure cabin, brought about by fatigue."

The solution to the Comet mystery, like the findings of most air disaster investigations, produced more than one culprit and no single cause. The first factor leading to the Comet crashes was a simply inadequate test program that failed to understand the long-range effects of continued pressurization and depressurization on airframe integrity. Compounding this error was the use of square-shaped window frames (Fig. 13) and the failure to incorporate mechanisms to prevent a fatigue crack from spreading into the aircraft's design. The third fatal factor was the fuselage skin—so thin that one U.S. airline president, on an early Comet demonstration ride, swore he could see the sides of the cabin walls "moving in and out like an accordion."

Impact

Three versions of the Comet were built—Models 1, 2, and 3—but only the MK-1 had been involved in the two Mediterranean crashes. MK-2 Comets, a larger and more powerful jet carrying up to 44 passengers, were extensively modified after the RAE investigation and used by a number of European airlines as well as the Royal Canadian Air Force. But de Havilland built only one Comet 3; the lingering stigma of the 1954 disasters, plus the advent of new, larger American-built jetliners, proved fatal to the Comet's viability.

Many aviation historians have assumed that the designers of the Boeing 707 and Douglas DC-8 learned valuable lessons from the Comet's mistakes, prompting them to design safeguards against metal

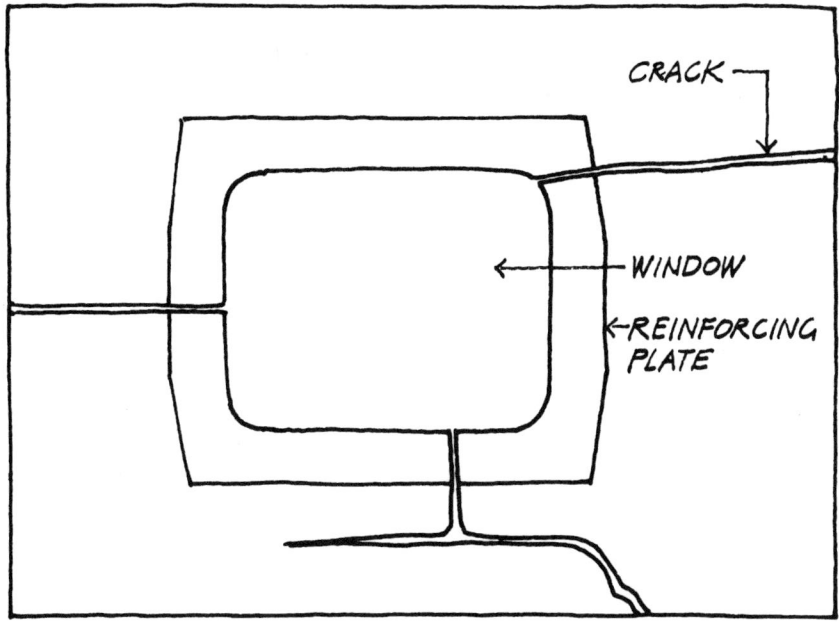

Fig. 13 Metal fatigue cracks around a small window in the upper fuselage developed during pressurization and depressurization cycles. These cracks resulted in disasterous structural failure of the pressure cabin.

fatigue and explosive decompression. In truth, more was learned from the Comet's pair of takeoff accidents than from Elba and Stromboli. The lesson learned from the several Comet crashes was the absolute necessity of flying jet transports "by the book," rather than the need to carefully test a new aircraft's design. Even before those tragedies, both Boeing and Douglas were already designing jet fuselages with thicker skin; triple-strength, rounded windows to distribute pressurization stresses equally around the frames; and metal bracing to provide additional structural strength (similar to reinforcing a wooden barrel with iron staves). The most important insurance against catastrophic decompression, however, were the small metal tabs or "stoppers" strategically placed throughout the fuselage so that if a fatigue crack should develop, its propagation path would be blocked before explosive decompression could occur. All rebuilt Comet 2s and the single MK-3 had such structural reinforcements as well.

Boeing dramatically demonstrated the efficiency of this preventative system before the 707 ever flew. Engineers deliberately weakened a 707 test fuselage with saw cuts up to 56 cm (22 in.) long, then pressurized the cabin and dropped five huge, stainless steel blades that slashed into the top of the fuselage. Small puffs of air excaped from the wounds, but there was no catastrophic explosion.

As a contrast to this successful demonstration, however, Boeing had first shown airline officials a Comet-like pressurized cabin without stoppers and dropped two of the steel blades onto the fuselage. A slow-motion film of the demonstration showed the metal fuselage skin beginning to curl outward at the points of penetration, then curling faster until the entire cabin split open and spit out its contents—seats, dummy bodies, and even the cabin floor. It was a reenactment of the Comet explosions.

SELECTED REFERENCES

- D. Anderton, "RAE Engineers Solve Comet Mystery," *Aviation Week,* Feb 7, 1955, p 28–42
- D. Anderton, "How the Comet Mystery Was Solved—Part II: Giant 'Jigsaw Puzzle' Gives Final Clue," *Aviation Week,* Feb 14, 1955, p 26–39
- "Comet Crashes: Dakar Accident," *Aviation Week,* July 27, 1953, p 16
- "Comet Crash off Elba Attributed by British Inquiry to Metal Stress," *New York Times,* Oct 20, 1954, p 14
- "Death of the Comet," *Time,* April 19, 1954, p 31–32
- D.D. Dempster, *The Tale of the Comet,* David McKay, 1958
- S. Hull, "Fatigue Blamed in Comet Crashes," *Aviation Week,* Oct 25, 1954, p 17–18
- N. McKitterick, "Comet Crash," *Aviation Week,* May 11, 1953, p 17
- N. McKitterick, "Comet Future?" *Aviation Week,* Jan 19, 1954, p 16–17
- "What's Wrong with the Comet?" *Business Week,* April 17, 1954, p 27–28
- "Why the Jets' Skin ruptured," *Business Week,* Nov 6, 1954, p 121–124

Case History Discussion

The case history included with this chapter reviews a failure that occurred in 1954, when commercial jet aircraft travel was in its infancy. At the time, many people feared the idea of commercial jet travel. In fact, a 1951 movie called *No Highway in the Sky,* based on the novel *No Highway* by Nevil Shute, played on those fears.

In the movie, three British commercial airliners go down for unknown reasons. The main character, played by Jimmy Stewart, predicts that the tail sections fall off after a certain number of hours and conducts fatigue tests to determine the moment of failure. In the end, of course, he is correct. It is one of the few movies in which fatigue testing plays a starring role!

Strangely enough, several years later the story came true. The British Overseas Airways Corporation (BOAC) Comet—the first commercial trans-Atlantic jet aircraft—twice broke apart in midair. The physical cause of the failure was cracks in the corners of the square windows. The initial flaws were small enough to remain hidden beneath the fastener heads, and, during operation, the cracks grew. The result was that the plane disintegrated.

What was the root cause of the failure? Was it the poor design of the square windows, which created a stress concentration? Was it the manufacturing decision that put fastener holes in the area of high stress concentration? Or was this decision based on schedule? A History Channel program on the Comet reported that the original design did not have fasteners in the corners. However, schedule pressure forced the chief design engineer to abandon the original design concept, which was very difficult to manufacture.

The ramifications of the failures eventually were catastrophic to BOAC. A competitor, the Boeing Company, was bringing out the 707. By the time BOAC got its fleet flying again, the 707 had taken over the skies.

Interestingly, one of the improvements that Boeing made during testing of the 707 was to increase the fuselage thickness. During full-scale testing of pressurized fuselages, the Boeing engineers took a fuselage like that of the Comet and dropped a ram into it. The fuselage basically blew up due to the pressure. When Boeing performed the same test on their thicker fuselage, it ruptured but did not explode. That basically sunk the Comet, which was shown to be an airplane designed to burst.

CHAPTER **2**

Failures Come in All Shapes and Sizes

IMPORTANT ASPECTS of failure investigation addressed in this chapter include:

- Failures come in all shapes and sizes
- Some failures you can touch and some you cannot
- Understanding what defines a failure
- Understanding why failures happen

Introduction

A failure analysis expert often is handed a failed component and then asked the simple question, "What caused this failure?" Performing a metallurgical evaluation may reveal fatigue striations or corrosion as the physical root cause. However, that may not be the reason the failure occurred.

Most failures are a cascade of events. For example, a failure may start as a stress corrosion crack, then turn into a fatigue crack, which eventually results in overload. That's three different types of failures rolled into one. Now back up even further. What initiated the stress corrosion cracking? Could it be management's decision to place the component in an environment for which it was not designed? In a true failure investigation, the hard part is to determine the extent of the cascading effect.

Many companies maintain metallurgical laboratories staffed by very talented people. But often these people simply perform metallurgical evaluations that discover nothing more than the physical root cause of a failure. Their thinking stops at a certain point.

As the person responsible for running a failure investigation, it is your job to learn more. You have to check what the metallurgical evaluators are doing and understand what they are doing, but you must also look at the bigger picture.

Some Failures You Can Touch, and Some You Cannot

Failures can involve individual parts, entire machines, or a process. Break down failures further and you will find that their origin can be physical (metallurgical mechanism), human (paper, error, sabotage), or latent (thinking, cultural). These possibilities open up the very concept of a failure, introducing things you can touch and things you cannot. The broken crankshaft illustrated earlier (see Fig. 5 in Chapter 1) is an obvious failure. The engine quit working immediately and, upon disassembly, the two pieces of the broken shaft provided the reason. Metallurgical examination of the crankshaft will uncover the failure mechanism.

However, consider a situation where a specification's poorly written requirements fail to discern the intended result. For example, specifications for many alloy systems contain various heat treatments that will result in various levels of mechanical properties. Each heat treatment is designed to produce a specific level of material strength and ductility. These specifications usually require a tensile test and a hardness test to provide evidence that the heat treatment was performed adequately. To reduce costs, some companies drop the requirement for the tensile test and only require the hardness test.

I once reviewed the heat treatment specification for a company that regularly heat treated Inconel 718, a nickel-base superalloy. In the specification the company listed five very different heat treatments designed to produce five very different strength levels. Each of these strengths had an associated set of mechanical properties: stress rupture, fatigue, fracture toughness, and so forth. Over the years the company had dropped the tensile testing requirement and relied solely on hardness testing. Unfortunately, the minimum Rockwell hardness level was exactly the same for all five heat treatments. Thus, using hardness tests alone, it was impossible to tell if the material had received the correct heat treatment. This situation had been going unnoticed for years. Where is the failure in this case? Was the failure the decision to save money by eliminating the tensile tests? Or was the failure the fact that no one technically assessed the specification to see if the hardness test was adequate on its own?

In Chapter 1, the levels of failure cause were grouped as:

- *Root:* the true cause of failure, encompassing the next three
- *Physical:* the failure mechanism (fatigue, overload, corrosion, etc.)
- *Human:* the human factors that lead to the physical cause
- *Latent:* the cultural/organizational rules that lead to the human cause

The broken crankshaft was an example of the physical level of failure cause. The specification with the poorly written requirements and possibly no subsequent technical assessment are two examples of the human level of failure cause. The company's culture of cost reduction, which led to removing the tensile testing from the specification, is the latent level of failure cause. So what is the root cause level? This is usually considered the true cause of the failure, the cause that prompts the cascade. Unfortunately, many failure investigations never reach root cause, but instead stop at the physical cause. The failure investigation team must decide how far the investigation will extend in order to determine whether human or latent causes exist as well.

Each Failure Is Unique. This fact is sometimes hard to remember when a particular failure looks just like one that was investigated previously. Resist the temptation to immediately assign the same root cause to failures that appear similar. If you fall into that mindset, you will give the same recommendation—which obviously did not solve the problem the first time.

For example, 16 years ago I was given a seven-ply bellows made from Haynes 188, a cobalt-base superalloy, for metallurgical analysis. The bellows was one of 15 in a heat treat lot that had been annealed at 1150 °C (2100 °F) for one hour. Unfortunately, the supplier had created an aluminum identification (ID) tag for each bellows, and during the annealing cycle each tag had been placed atop the corresponding bellows. Aluminum melts at approximately 650 °C (1200 °F), so the tags melted during the 1150 °C (2100 °F) annealing heat treatment and began to erode through the bellows plies. Because these parts were expensive, my assignment was to perform a destructive metallurgical analysis on one bellows in order determine how many plies were still intact so that the other bellows from the same heat treat lot might possibly be used under reduced operating conditions. Testing showed that too many of the plies had been eroded through, and consequently the bellows were deemed unacceptable for use. My recommendation to the supplier was to use stainless steel ID tags, which would survive the annealing cycle.

I considered this failure to be unique and having very little chance of ever happening again. I was wrong. About nine months later, the same type of bellows was presented for another metallurgical analysis. The supplier had changed the ID tags to stainless steel, but had neglected to change the wire that attached the tag to the bellows. In another lot of bellows annealed at 1150 °C (2100 °F), the aluminum wire had started to erode through the plies. The project leader wanted to cancel the metallurgical analysis and scrap the bellows based on the previous failure. His argument was that it was the same failure and would have the same outcome, so why waste time and money. My manager and I recommended the metallurgical analysis be completed since the effort was minimal compared to the cost of the bellows. The analysis showed that the aluminum wire had eroded through only one ply, and it was determined the bellows were acceptable for use.

Interestingly, about four years ago, I was given a fastener fabricated from A286, an iron-base heat-resistant superalloy, that had been solution heat treated at 885 °C (1625 °F) with an aluminum tag and wire attached. Once again, the aluminum melted during the heat treatment and started to erode into the A286. My first reaction was amazement that I would see this type of failure once again. After considering the differences and similarities between Haynes 188 and A286, my first thought was just to scrap the parts and save the time and money of the metallurgical analysis. But I remembered what had happened with the bellows. Based on the analysis results, we were able to save 50% of the fasteners.

Failures Are Specific to an Industry. Another good point to remember is that failures are specific to an industry and the specific requirements of that industry. Always make sure you know the defining requirements of the failure at hand. The saying "One man's trash is another man's treasure" also applies to failures.

For example, thin dense chrome plating on AISI 52100 carbon steel bearings without a 190 °C (375 °F) bake to remove the hydrogen absorbed during the plating process has been used in commercial applications with great success when replacing nonplated AISI 52100 carbon steel bearings. Inexpensive AISI 52100 carbon steel is the bearing material of choice for bicycles, cars, trucks, and numerous other applications. It contains 1% C, which makes it very hard and strong, a perfect bearing material, but it is not corrosion resistant.

A company came up with the idea of using a low-cost method of making AISI 52100 corrosion resistant by applying thin dense chrome plating to its AISI 52100 bearings, a process that has been around since the early 1980s. The bearing life of the plated bearings, especially in applications that had aggressive and wet environments, was greatly improved. Changing from the AISI 52100 carbon steel bearings to the same bearings with a thin dense chrome plating increased the life of the bearings. The company was a hero to its customers. In addition, the AISI 52100 carbon steel with the chrome plating process and no 190 °C (375 °F) bake to remove hydrogen was also cheaper than using AISI 440C stainless steel bearings. It proved to be such a financial success that the bearing company began to search out other opportunities and other markets.

One opportunity the company identified was using the chrome-plated AISI 52100 bearings to replace AISI 440C stainless steel bearings in aerospace applications. AISI 440C stainless steel is the bearing material of choice for the aerospace industry because it is a corrosion-resistant steel and thus satisfies a basic industry requirement. Once the chrome-plated AISI 52100 bearing component moved into aerospace applications as a low cost replacement for AISI 440C stainless steel, problems arose. The service requirements were different for aerospace applications, and the lack of a 190 °C (375 °F) bake to remove hydrogen generated during the plating process became a liability. In the aerospace industry, bearings operate in less aggressive environments and the bearing life depends more on the bearing material than on the environment. The lack of the 190 °C (375 °F) bake to remove hydrogen is more important to the bearing life when considering thousands of hours instead of hundreds of hours in operation. Lastly, the aerospace industry requires the 190 °C (375 °F) bake to remove hydrogen after plating operations—it is not just a common operation, but a required one. The company did not understand the technical reasons why the 190 °C (375 °F) bake process was required or that the aerospace industry requires the process. The bearing company had to spend many personnel hours justifying the elimination of the 190 °C (375 °F) bake to remove absorbed hydrogen, which drained their financial margins.

In this situation the company was a hero in one industry, but a failure in another because it did not understand the requirements of the new industry it was trying to enter.

Another example is the short precipitation-age heat treatment used on nickel-base superalloys developed for oil field applications. The oil field industry requires drill bits to dig into the earth under harsh conditions. These drill bits must be hard for good digging capability and wear life, but also tough to prevent fracture. These two material properties—hardness and toughness—are inversely related, but since fracture is more damaging to the drilling system, the drill bits are designed for the best wear life they can achieve without fracture. Every time a drill bit breaks, it takes hours to get it out of the hole and replace it. Downtime is lost money. The oil field industry had been using low-alloy steel drill bits heat treated to provide the best compromise between the two required properties.

In the early 1980s, however, the oil field industry started using nickel-base superalloys for drill bits because these materials demonstrated improved wear life characteristics and improved fracture toughness. Because of their good strength and fracture toughness, nickel-base superalloys have long been the material of choice for jet engines and space vehicles. The materials are well documented, with plenty of data for long-life applications.

Unfortunately, nickel-base superalloys cost about 10 to 15 times more than low-alloy steels. They also required an expensive twenty-hour precipitation-aging heat treatment cycle, developed by the companies that produce jet engines and rocket engines in order to provide the best mechanical properties (including fatigue life) for their applications. In response, the oil field industry developed a three-hour precipitation-aging heat treatment that provided adequate fracture toughness until the drill bit wore out. In this industry, the twenty-hour heat treatment was unnecessary.

When the oil field industry slowed down, this short precipitation-aging heat treatment was introduced by some heat treat suppliers into aircraft and aerospace applications as a cost reduction concept. Unfortunately, very little data existed on long-life mechanical properties, such as fatigue life and stress-corrosion cracking, that are necessary to the aircraft and aerospace industries. In one industry these heat treat suppliers are a hero, but in another industry with different expectations, they were a failure. Marine versus oil field versus automobile versus aerospace—even though a concept works in one industry, it may not work in another.

As a failure analyst, you must always remember, business is business. People keep trying to find market share somewhere else, and they will take what they know into new markets whether or not they understand the business requirements.

What Defines a Failure?

In very broad terms, failures can be divided into two categories. The first category encompasses the most common concept of failure: A component, machine, or process fails, and everything stops. Failure of a component like a crankshaft, bearing, or bracket is quite obvious, as is failure of a machine—be it a commercial airplane or a nutcracker. Determination of the physical, human, latent, and root causes of the failure usually is achieved. The difficulty of determining the failure's root cause increases with the complexity of the component or machine, but it generally can be discovered.

Failure of processes such as heat treatment, coating, plating, welding, mail delivery, airline on-time arrival, or parking lot arrangement can be more complex, as is determination of the root cause. In addition, the cause-and-effect relationship or the cascading mechanism for the failure of processes may not be readily apparent.

It is interesting to note that the general public does not readily accept the failure of components or machines; however, the failure of processes such as mail delivery, airline on-time arrival, and food service at restaurants is acceptable to a point. The point at which the failure of these processes is unacceptable is a moving target that varies drastically not only with the individual but also with the type of day the individual is having. For example, on a recent trip my flight was an hour late leaving the airport, with a scheduled fifty-minute layover. Therefore, I shouldn't have been able to make my connection, but I did. I did not want an explanation, because as long as I caught my connecting flight, I didn't care. We all put up with these kinds of process failures every day, ignoring them as long as our lives are not affected.

In the second failure category, a component, machine, or process fails to achieve performance criteria such as operating life, operating limits, and specification requirements. People hold expectations for the life of items such as car engines, tires, household appliances, biotech implants, and even pens. These life criteria are usually not written down, but are debated in courtrooms. If a set of car tires lasted 48,000 km (30,000 miles), most of us would be happy, but if the tires lasted only 16,000 km (10,000 miles) you would be speaking with the tire dealer.

Biotech implants are an interesting example. Originally, knee and hip implants were designed for patients in their 60s who led a reserved, nonactive life. The idea was to design a permanent implant capable of providing a normal lifestyle prior to death. Research concentrated on how to make bone grow into the implants for a more secure fixture. Today, with the general population living longer and, more importantly, staying active longer, implants are being placed in much younger patients—in some cases, pro athletes. These implants are expected to last longer and perform at much higher levels. In addition, people want to replace the implants as they wear out, which was not part of the original design.

More specific than operating life are operating limits, such as gas consumption, gas flow, production scrap rate, modem speed, specification turnaround time, computer chip speed, or acquisition frequency. These limits are usually well defined and in most cases are mutually agreed upon by the parties involved.

An example of specification failure involves the suspended laser acquisition pods (SLAPs) that pop up on military helicopters to "paint" a target. Once the helicopter is in position, planes come in with bombs or ships fire missiles, and the ordnance follows the tracking of the acquisition pod. A problem arose, however. The error band designed into the pod and the error band designed into the missiles were not consistent. The two error bands could actually fall out of each other, resulting in the SLAP unit being incapable of performing its task—a complete failure of the system.

At first, the engineers examined everything individually. The SLAP unit met all of its specifications. The bombs and missiles met all of their specifications. Although everything was operating to very closely specified requirements, the system would not work. Finally, one engineer looked at the complete system and compared the various individual specifications. The error band of each of the units was changed to fall within the other units' operating range, and the problem was solved. This is an example of a failure of the system even though there was no failure of the individual components.

More recent examples of failure due to unsatisfied specifications include the two 1999 missions to Mars where one group of scientists was discovered to be working in metric units while another group was working in English units. Both vehicles were lost due to this misreading of the mission specifications.

Lastly, specification requirements such as mechanical properties, plating thickness, weight, and coating optical properties are very well defined in written documentation. The specific definitions or limits usually are created by the customer, with tests required to ensure compliance.

In this section we moved from defining a failure via personal impressions or opinions, to defined criteria, and finally to very well-defined criteria. Specifications for properties such as hardness or tensile strength represent very well-defined criteria. In such cases, tests are performed to confirm that the specification requirements are being satisfied, and failure becomes very obvious.

Why Do Failures Happen?

This question has been asked many times. Over the past 50 years or more, the reasons for failure, as well as their order of priority, have changed. The 1948 edition of the ASM *Metals Handbook* listed the following three categories as the reasons for failure:

- Design factors
- Processing
- Service environment

Interestingly, the 1948 *Metals Handbook* assumed proper material selection if the engineer followed the criteria set forth in the handbook.

In the 1961 and 1975 Republic Steel Corporation handbooks on failure analysis, these six categories were given as the reasons for failure:

- Design
- Steel selection
- Heat treatment
- Material quality
- Method of fabrication
- Assembly

Republic Steel, of course, was oriented toward steel products. Therefore, it is interesting to note that proper material, or steel, selection is second on this list, but was absent in the 1948 list. Also note that service environment does not appear in this list at all.

I currently use the following list:

- Service or operation (use and misuse)
- Improper maintenance (intentional and unintentional)
- Improper testing
- Assembly errors
- Fabrication/manufacturing errors
- Design errors (design, stress, material selection, etc.)

Service or Operation. Today, service or operation is the first suspect when a failure occurs. As noted earlier, most components and machines, even processes, have a certain life expectancy. Under normal use most items will wear out in an acceptable time frame. A failure occurs when an item wears out sooner than the user expects. At this point, however, it is important to distinguish between normal use and misuse. A pencil, when used in a normal manner, has an expected life. Misuse such as chewing on the pencil or breaking it in half to share with a fellow student reduces the pencil's life.

Cars provide another good example. Most cars will last many years if maintained and driven correctly. In the early 1990s car manufacturers began offering six- to ten-year warranties on new cars, which gives some indication of how long they expected the car to last. However, constantly driving a car at high speeds or on rough roads will reduce its life and perhaps void the warranty. When I was in high school during East Coast winters it was considered fun to drive cars into snowbanks or spin "donuts" on slick, snow-covered parking lots. Unfortunately, it was not unusual to find a street sign, small tree, or fire hydrant buried in the snowbank. It was also easy to lose control during a spin and slide off the parking lot into deeper snow. Objects such as cars are not meant to be used in certain ways.

Tools are regularly misused. A simple screwdriver may be used as a weeding tool, a wedge, a pry bar, or even a weapon. A few of my high school friends worked in a garage and were instructed to use expensive Craftsman screwdrivers as alignment tools when mounting a tire onto a truck or car. The rationale for using an expensive screwdriver instead of a cheap one was that if the tire slipped and sheared the screwdriver, the Craftsman warranty guaranteed the garage a new screwdriver. Screwdrivers are not meant to be misused, but who cares if you can get a new one?

Misuse also extends to larger tools. Although machining equipment and tooling is designed to run at a certain capacity, oftentimes it is run at 110%, or more, of rated capacity. The owners are relying on what they believe is the inherent overdesign of machines by the company that produces them. That constitutes misuse, but if you are getting more parts per hour, then is it worth losing one or two machines?

The infamous yearly list of "Darwin Award" winners is a good place to find examples of the obvious misuse of machines. Past honorees include two men who used a lawnmower to cut hedges by lifting it into the air and holding it. Obviously, this idea produced an accident. Reportedly, the two men successfully sued the lawnmower company on the grounds that they had not been told not to operate the equipment in that manner. Manufacturers now include a label warning users to keep hands and feet away from lawnmower blades during operation.

Improper maintenance is the second reason for failures. Home appliances, computers, cars, and aircraft all should be properly maintained, but often are not. Anyone who reads the news knows that airlines sometimes neglect airplane maintenance and mechanics sometimes improperly maintain cars brought to their shops for service. When was the last time you changed your car's radiator fluid or rotated its tires? When was the last time you changed the filter in your home ventilation system or drained the water heater?

Improper maintenance results in life reduction or complete breakdown. Sometimes it is unintentional, as in the case of the car owner who decides that if a 50/50 ratio of radiator fluid to water is good, then 100% radiator fluid must be better. That may sound like a good idea, but in reality undiluted radiator fluid is a very aggressive solution and erodes the radiator. Other times improper maintenance is intentional, such as skipping the overhaul of a machine in order to squeeze out a few more weeks of use.

For example, I once worked in the program office for the maintenance of commercial airliner jet engine nacelles (Fig. 1). A nacelle is the structure that surrounds the jet engine and contains the fuel and electrical connections, the structural interface, and some structural hardware that includes the thrust reverser used during landing to reverse airflow forward and provide additional braking. One day, the company began receiving requests from a small third world airline for more brake pads for its fleet of planes. These requests began to arrive at an ever-increasing rate. The program manager finally sent an engineer to look into the problem. The engineer reported that what was needed was not more brake pads, but new thrust reverser doors. The real failure was that the airline was breaking the thrust reversers. The movement characteristics of the doors were wear-

Fig. 1 Jet engine nacelle on a commercial airliner

ing out, but instead of replacing them, the airline mechanics bolted them down and immobilized the thrust reversers—rendering them essentially useless when the planes landed. The pilots had to step down hard on the brakes to stop the planes, thus quickly wearing out the brake pads. The difference was that the brake pads had a warranty if they wore out too soon, but the thrust reversers did not. Therefore, the airline could get the brake pads for free, but fixing the thrust reversers cost money. In the end, it took someone looking at the problem firsthand to discover the actual failure root cause.

A common outcome of poor maintenance is corrosion. In a 1998 study sponsored by the U.S. Department of Transportation and NACE International, the Corrosion Society, the direct cost of corrosion in the United States alone was estimated to be $276 billion each year, and this loss frequently is due to improper maintenance (Ref 1). Unfortunately, the cost of proper maintenance can also be quite high. The Golden Gate Bridge is a world-renowned landmark that connects San Francisco with smaller cities to the north. Maintenance crews work on the bridge year-round to stay ahead of corrosion problems. The crews strip the paint and corrosion, then repaint the bridge. They start at one end, complete the entire bridge, and then start all over again. The bridge is being stripped and painted every day of the year, every year. This work is done for two reasons. The first is to prevent structural damage and extend the life of the bridge. The second is to make the bridge aesthetically pleasing. As one of the most photographed structures in the world, the Golden Gate Bridge is a symbol of San Francisco. The continual maintenance is a matter of civic pride as well as life extension.

Improper Testing. The third reason for failures is improper testing. Today's prevalent cost-driven, "hire and fire" business environment often means that test-related work is in the hands of the lowest-cost personnel capable of performing the task. Sometimes these workers have just enough training and sometimes too little training. Unfortunately, this business attitude can foster a situation where such personnel may cause a failure by selecting the wrong test, performing the test incorrectly, or reviewing the test results incorrectly.

Here's an example of improper test selection. While inspecting a part using dye penetrant, a supplier discovered a surface flaw. The flaw failed the acceptance criteria, but was different from any other flaw the supplier had observed before. The supplier drew up the proper paperwork to document the inspection and submitted the part and paperwork to his customer for evaluation. The first operation at the customer was to reinspect the part and confirm the existence and position of all flaws. However, the dye penetrant inspection area was too busy, so the planner decided to use magnetic particle inspection. Both processes are excellent methods for detecting surface flaws. However, one requirement for magnetic particle inspection is the part itself must be magnetic. This was not the case. The part passed the inspection and the customer sent it back to the supplier with a statement that there was no flaw and instructions to complete the part. The supplier reinspected the part using dye penetrant, discovered the flaws again, and called the customer. Many wasted labor-hours and lost schedule days later, it was discovered that the wrong inspection process had been used to evaluate the part. The supplier once more sent the part to the customer for evaluation.

A classic example of incorrect test performance concerns bird strike testing on jet engines. Bird strikes have been a problem during aircraft take-offs and landings for many years. A bird may even be ingested into an engine at take-off or landing. The Federal Aviation Administration (FAA) requires that an engine must be able to handle a bird strike without uncontained failure, fire, or engine mount failure. The engine must be able to shut down in a controlled fashion to pass this FAA requirement. The jet engine industry actually has a test to determine the effect of a bird strike on an engine. Frozen chickens are thawed and shot from a cannon into a jet engine running on a test stand to determine if the engine will stay within its nacelle. One day a failure was reported at the test stand when one of the engines exploded during the bird strike test. The technician shot the chicken into the engine, and the engine came to pieces—a very rare event. During subsequent review it was discovered that the new technician who performed the test had apparently skipped the step about first thawing the chicken. Needless to say, the engine failed the test when the frozen bird hit it with the destructive impact of a bowling ball. Low-cost personnel often equals clueless or untrained personnel. It was a simple test: Thaw a chicken and shoot the cannon.

Lastly, reviewing test data incorrectly can be costly. As a point of reference, in many large companies the cost just to create the paperwork to evaluate an unacceptable condition discovered in hardware is $2000 to $5000. Then the cost of the engineering evaluation must be added.

One day an engineer approached me with the paperwork generated for an unacceptable condition on an aluminum die forging. The forging had been tested in three directions to meet the drawing requirements: longitudinal, long transverse, and short transverse. The short transverse direction had passed, but the longitudinal and long transverse directions had failed. I looked at the data and noticed something very odd. The short transverse tensile strength data was higher than both the longitudinal and long transverse tensile strength data. Normally the short transverse tensile strength data is the lowest of the three directions. I told the engineer to call the supplier and check the data. I believed the test data had been mistakenly transposed when written on the report sheet.

The supplier confirmed my theory and corrected the report sheet. The forging was acceptable. The company had just generated a $5000 piece of paper, and the engineer had spent hours on the evaluation by reviewing the documents and the drawings and familiarizing himself with the hardware so he could explain it to me. Thousands of dollars were spent in engineering evaluation because the quality-control personnel did not have enough experience or training to look at the data and realize the numbers had been transposed.

These are just a few examples of the wrong personnel doing the work. Wrong in the sense that they were untrained, unsupervised, incapable, or a combination of all three. In today's cost-conscious business world, the current philosophy is to have the lowest-cost personnel do the work, not the personnel who can do it well. Consider the fact that in the end, it may cost the company more.

Spend enough time around component testing and test areas and you will learn that all potential causes of failures or events are taken seriously, no matter how ridiculous they may seem. I was working in a company that produced rocket engines. Part of that business consisted of regularly performing rocket engine tests. One time a series of problems with the testing occurred within a very short period. I was present to support one of the next tests and noticed a test engineer sitting in the back of the test bunker. I said hello and asked how he was doing. He answered that he was in a lousy mood because he'd been told he would not be allowed to start the test that day. I asked why and he said that someone had noticed that he had started 75% of the tests that had had problems in recent weeks. Therefore, he was now banned from starting any tests. The theory sounds good enough until you consider the fact that all that was required to start one of these tests was to push a button. Statistically, they were right, but realistically, this had nothing to do with the problem. This is how crazy things had gotten with the testing. The root cause of the test anomalies was discovered and, of course, it was determined that whoever pushed the start button did not contribute to the problem.

Assembly Errors. The fourth reason for failures is assembly errors. Today's manufacturing world produces products faster and cheaper, but are they better? Sometimes speed does kill.

Before September 11, 2001, the Boeing Company produced 500 to 600 aircraft each year. Their goal was to produce an aircraft per day on the 737 production line. The actual production time at the Boeing assembly plant for a 737 aircraft is three to five months. The assembly plant goal is to collapse that time frame, which is the goal of any assembly plant. But when is fast too fast? When do assembly errors begin to occur?

Car companies in Japan can produce a car every twenty seconds, and aircraft companies work toward producing a commercial jetliner every day. It is an amazing feat, but the combination of fast and cheap with the use of low-cost personnel can be disastrous. Design and manufacturing engineers believe they can control the problem by making products and fabrication processes "idiot proof." However, time and time again failures occur that prove the "idiots" are smarter than the engineers think.

The high-pressure fuel and oxidizer turbopumps on the Space Shuttle main engine (SSME) rotate at approximately 35,000 and 28,000 rpm, respectively, and develop 69,000 and 25,000 hp, respectively (Fig. 2). These two pieces of machinery are attached to an engine that is 4.27 m (14 ft) tall and 2.29 m (7.5 ft) in diameter at the exit nozzle, weighing 3402 kg (7500 lb). The turbopumps are two of the highest power-density machines on earth.

Assembly of these turbopumps is a very well-designed and controlled engineering feat, but errors can occur. The high-pressure fuel turbopump has two turbine discs that are assembled back to back with fasteners. It is a critical assembly and the design engineers decided that in order to ensure that the two turbine discs were assembled correctly, they would put a pin in one turbine disc with a matching hole in the opposite disc. If the two discs were not assembled correctly, then the pin would cause an obvious interference and be noticed. A design engineer told me it was completely "idiot proof." From a design standpoint, putting a hole in a turbine disc that spins at 35,000 rpm can create a notch effect and increase local stress,

Fig. 2 Space Shuttle main engine

a fact that drove the stress engineers crazy. However, the design engineers prevailed because of the need to ensure proper assembly.

During testing of an SSME one day, the monitoring devices noted that the fuel turbopump was not operating correctly. The test sequence had to be stopped and the fuel turbopump removed from the engine while it was still on the test stand—a costly and time-consuming process that delayed delivery of this $60 million product.

The fuel turbopump was returned to the assembly facility where the design team had gathered for the disassembly of the turbine end. I will never forget the look on the faces of the design and manufacturing engineers when the turbine disc stack was removed. The turbine discs were found to be stacked upside down, an obvious assembly error that the engineers all agreed was impossible to accomplish—in theory, at least.

A more common example of assembly errors involves truck radiator fans. Many Californians enjoy getaway water-ski weekends to the Colorado River. Large pickup trucks or SUVs are popular for hauling ski boats through the desert between California and Colorado, where temperatures routinely reach more than 40 °C (100 °F).

A friend of mine had purchased a new truck for this very purpose. During the first trip to Colorado, the truck began to overheat and he had to stop and wait for the engine to cool down a number of times. On his return, he went to the dealer and told them there was something wrong with the truck. They checked the vehicle from grille to bumper and could find nothing wrong. They suggested that he add the cooling system upgrade since he was hauling such a boat across the desert. He did. During the next trip, the same overheating problem occurred.

The dealer checked the truck once more and again found nothing wrong. My friend was relating this story to a few of us outside his home. The truck was idling as we stood around the front grille. Someone dropped a napkin and, as it flew in front of the truck, the napkin was blown forward away from the truck. A couple of us were puzzled. I picked up the napkin and brought it back to the grille. As we watched, the engine fan blew the napkin away. In order to cool the engine, the fan should have been sucking air *into* the grille and blowing it over the engine. The huge fan used to cool the truck engine was on backwards. It had been assembled wrong and inspected twice, but the error was never discovered. We suggested that our friend go back to the dealer and have the fan reversed. Now the temperature gauge on his truck never leaves the cool zone. How difficult is it to install a radiator fan? Doing so improperly leads to huge ramifications for the user.

I used to support the assembly line for older transport aircraft nacelles in San Diego. The assembly line consisted of seven men who had been together for 15 or more years. The company decided to move the assembly to a new facility in Alabama, where labor was cheaper. All the drawings were sent to Alabama to set up the assembly line. But a small problem arose. Workers there could not make a single nacelle using the drawings provided. Engineers were sent to the assembly line in San Diego, which was still operating while the Alabama plant got up to speed. The engineers watched for a few days and discovered that the workers in San Diego had been reengineering the nacelle among themselves without telling anyone for 15 years. They implemented very small changes, mostly within the tolerances of the drawing, that made the nacelle assembly work. For example, the man in station 6 would be having trouble assembling a certain part. He would go down to station 1 and talk to the man there about moving another part just a little bit.

The engineers had to sit down, starting at station 1, and find out from each worker what he was doing. This resulted in more than 700 changes to the drawing. These men had been assembling the nacelles by working out problems among themselves. Parts that came back for maintenance came to them, so they knew what to do. But when the nacelles were sent to Alabama and the assembly line tried to use the existing drawings, they did not stand a chance.

Where is the failure in this case? Was the latent cause the fact that engineering procedure should not have allowed the manufacturing personnel to deviate from a drawing? Was the human cause the fact that engineering had become too distant from an older aircraft line and did not respond to requests for assistance, so the production personnel fixed it themselves? In a production environment, that information must get back to design so there is documentation of what happened. In every instance there should have been a request for a design review and a drawing change. But that takes time and money. And these men thought, it is just a little thing, I am just tweaking this a little. But over time there were so many tweaks that the drawings were no longer valid.

The automobile industry is always looking to reduce costs. Many company programs provide incentive to assembly personnel who suggest the removal of a single screw or fastener. That single component multiplied by thousands of units over multiple years will save the company a lot of money. The programs give the employee a one-time bonus or a percentage of the calculated cost savings.

One assembly-line employee suggested the removal of one bolt on a four-bolt pattern that held a small component to the engine. The company accepted the idea for evaluation and gave the employee an award certificate for the suggestion submission. Unfortunately, the employee interpreted the award as the acceptance of the idea and started to remove one bolt from the four-bolt pattern. Months later, cars began showing up at dealers with the component coming loose on the engine. There was only one assembly line, and they traced it back to the one employee.

The automobile company did not fire the worker. To its credit, the company determined that the root cause of the problem was the manner in which the suggestion program operated. What had happened was that the suggestion went to the design group, who evaluated it and then rejected it. But no one ever reported this back to the employee. The company changed the initial certificate to clearly indicate that the suggestion was under evaluation but had not been implemented. This is a good example of a successful failure investigation. They found the physical cause: a missing bolt. They found the cascade human cause: a worker who was not putting in the bolt. And they found the latent cause: a suggestion program that was a bit inadequate.

Lastly, it is getting more and more difficult to check for assembly errors because the products we make are becoming increasingly complicated. In addition, companies are using fewer employees and using the lowest-cost employee capable of performing the task. There is no such thing as idiot-proof assembly. Manufacturing personnel find a way to "make it work," often introducing errors in the process. More and more, companies are depending on automated processes to check for errors. But automated processes do not think or notice when "something just does not look right." They are capable only of go/no go decisions based on predetermined criteria. Automation is producing the product faster and inspecting it faster, but is it doing the job?

Communication is a cornerstone of any manufacturing operation. The following is an example of poor communication between two personnel who clean aircraft.

John is the chief airplane washer at the company hangar and he hooks a high-pressure hose up to the soapsuds machine. John then turns the machine "on." Shortly thereafter, he receives an important call and has to leave work. As he departs, he yells to Don, his assistant, "Don, turn it off." Assistant Don thinks he hears, "Don't turn it off." He shrugs, and leaves the area right after John. The result the next morning is a suds-filled hangar (Fig. 3).

Fabrication/Manufacturing Errors. Let's move on to the fifth reason for failures. Before a component can be assembled, its parts must be fab-

Fig. 3 Squeaky clean airplanes in a hangar filled with suds

ricated or manufactured. Fabrication/manufacturing errors are easier root causes to identify and are more commonly noted in the public domain through lawsuits and newspaper articles.

For example, an engineer in a forging company noticed a worker getting ready to load a pallet of aluminum billet stock into a furnace set at 980 °C (1800 °F). As noted previously, aluminum has a melting temperature of about 650 °C (1200 °F). The forging shop specialized in steel forgings, but was quoting work in other materials to expand the company into new markets. The shop personnel did not recognize the aluminum material as different from the normal steel material and was going to put the "steel" into the furnace. The engineer stopped him, though not without a lot of commotion caused by interfering with union personnel in the performance of their job. The engineer saved more than just the single job of aluminum forgings. As noted earlier, molten aluminum will erode through materials. It is fairly certain the molten aluminum would have eaten its way through the bottom of the furnace, resulting in millions of dollars in production loss, downtime, and furnace replacement.

Another example involves the SSME high-pressure oxidizer turbopump discussed earlier. A mechanic called me over to look at a turbopump that was being disassembled. One of the parts in the top portion of the turbopump that was used to cool the turbine disc had a series of cracks in it. The part was made from a 321 CRES tube in the shape of a ring with multiple small cooling jets welded into it. The 321 CRES was chosen because it was a stabilized stainless steel and not prone to cracking during welding.

A Shuttle launch was only four days away, and a problem like this would delay the launch due to the fear that it was not unique but system-wide on all turbopumps. I began researching welding defects in 321 CRES. About an hour later the director of SSME Turbomachinery walked up with the part in his hand and put it on my desk. He told me I had three days to get the answers to two separate questions before the launch had to be scrubbed. One, what was the technical reason the 321 CRES had cracked? And two, was the problem unique to this particular turbopump or was it systemic to the Shuttle fleet? One manner in which NASA evaluates a failure is whether it is a unique failure or a systemic failure. If you can demonstrate that a failure is unique, NASA will expect a solution. If a failure is shown to be systemic, however, they will ground the Shuttle fleet until the extent of the problem is determined and a solution presented. I worked for the next seventy-two hours straight coming up with the answers to those two questions.

The technical reason for the cracking was liquid metal embrittlement (LME) of the 321 CRES by copper. Liquid metal embrittlement is a specific failure mechanism that is under the more general category of environmentally assisted failures. Certain metals in their molten state embrittle other metals when they come in contact. The LME of 321 CRES by copper is a classic failure mechanism and well documented. Metallographic evaluation determined that the cracking was intergranular, and scanning electron microscopy revealed copper on the fracture surfaces of the cracks. There was only one small problem: Copper is not used in the turbine end of the high-pressure oxidizer turbopump. So where did the copper originate?

The flight was launched on schedule. The physical cause was presented to NASA and accepted. The problem was deemed not systemic to the fleet when information concerning the examination of other turbopumps noted that the same component was not cracked.

But the failure investigation was not complete. The physical cause had been determined, but the root cause had not. Therefore, a team was sent to the supplier who fabricated the part. The welding equipment used to fabricate the part had a grounding wire as part of the welding machine. The end of each grounding wire usually had a clip whose function was to clamp to the workbench. These clips are usually made from a copper alloy. Normal welding practice is to ground to the workbench, but it is not uncommon for a welder to ground to the part for convenience. It is also not uncommon for a spark to occur when clamping the grounding wire. This scenario had been previously proposed as one of the root causes in the fault tree for this failure investigation.

The supplier was to have been told that the purpose of the visit was simply to tour the facility and watch the welders at work. After our arrival, we were given a briefing on the company's history and then taken on a shop tour. As we approached the door to the welding area, we noticed a new sign over a small bucket that read "Deposit All Pennies Here." The shop manager told us it was to ensure that no copper was in the welding area. The team members looked at one another and smiled. The cat had been let out of the bag and the tour of the welding shop was just a formality at that point.

On the way home, we asked our supplier management contact if the supplier had been told the real reason behind the team's visit to the shop. He responded that the supplier had been told there was a problem involving copper. Why, was he not supposed to tell the supplier? The failure investigation team had had its own little failure of communication.

Even without direct proof, the team determined that the most probable root cause was a human cause: Someone had clamped the grounding wire from the welding equipment to the part and deposited molten copper due to a spark. Months later, our new company calendar was distributed. Each month featured a collage of photographs from suppliers. We were glancing through the photographs when someone told me to look at the bottom left corner of the October calendar page. There was a photograph of one of the welders from the shop that had worked on the failed part. The welding equipment was clearly grounded to the part. That was one of the few times I ever found absolute proof of the root cause of a failure.

Fabrication/manufacturing errors can come in many forms. Processes such as welding, plating, coating, and heat treating are routinely done by outside suppliers. In the past 15 years, many heat treat suppliers have reduced their number of engineers in order to cut costs. These suppliers think that if they produce adequate written instructions, anyone can heat treat hardware. Earlier in this chapter, an example was presented of a heat treat supplier who ruined two orders of bellows made from a cobalt-base superalloy by annealing the bellows at 1150 °C (2100 °F) first with aluminum tags and then again with aluminum wire. Perhaps an engineer would have thought to remove both the tag and the wire.

Another heat treating example involves the interpretation of heat treating specifications and knowledge of heat treat furnaces. For years the industry heat treat specification for steels has been MIL-H-6875. For use in the International Space Station (ISS), a part called a trunnion pin was made from Custom 455, a precipitation-hardening (PH) martensitic stainless steel, required heat treating to the H1000 condition. The first step in this process is to solution heat treat the part at approximately 830 °C (1525 °F), then quench in oil. The second step is to precipitation age the part at 540 °C (1000 °F) for four hours. The part in question was used to hold a large ISS assembly in the Space Shuttle's cargo bay for delivery into space.

Tensile test coupons were required to demonstrate that the heat treat supplier had heat treated the parts in an acceptable manner. Coupons representing two-thirds of the parts were found to be below specification minimums by 10 to 15%. Even the coupons that passed the test were barely above the specification minimums. Unfortunately, the parts were needed to meet schedule, so a failure investigation team was assembled.

A review of the heat treat supplier process records indicated that the coupons and parts had been solution heat treated in a vacuum furnace. However, the coupons and parts that had passed had spent 90 to 120 minutes in the furnace prior to quenching, whereas the failed coupons and parts had spent only 45 minutes in the furnace prior to the quench. The heat treat supplier stated that he had followed the specification. MIL-H-6875 required only 30 minutes minimum at the 830 °C (1525 °F) solution heat treat temperature and the thermocouples used to monitor the parts indicated that the requirement had been satisfied.

What the supplier failed to realize was that the specification was a requirement, but not an absolute requirement, and that some engineering judgment was necessary. The judgment in this case concerned the type of furnace used. Because there is no air to conduct heat to the parts, vacuum furnaces heat parts by radiation. Vacuum furnaces also can present a problem called "shadowing," where one part blocks other parts from the radiating heat, thus causing the shadowed parts to heat up at a much slower rate. The thermocouple on the parts during the solution heat treat may have indicated that the coupons and parts were at 830 °C (1525 °F) for an adequate period of time, but the tensile test data indicated the opposite. The supplier did not even believe his own information. The coupons and parts that had seen longer times, 90 to 120 minutes, had demonstrated improved tensile test data. In the end, the supplier agreed to extend the solution heat treat time to four hours. The tensile test results from subsequent coupons were exactly what was expected.

In the past, welding and heat treat shops were the experts and would tell their customers this type of information. Many companies relied on their suppliers for technical expertise. That is just not 100% true anymore. In order to cut costs, many suppliers no longer hire engineers with technical expertise, and instead depend on their customers to help with problems.

Machine shops can introduce a different type of manufacturing error. Drawing interpretation is a skill not taught in colleges to any detail or expertise. On-the-job training is how most engineers learn to read a drawing. However, drawings of intricate parts can be difficult to translate into three dimensions. Many parts are shipped with a feature or hole in the wrong location.

Today, computer files and three-dimensional models have replaced the blueprint drawing. Unfortunately, drawing interpretation is still learned on the job, although the expectation is that interpretation errors will be reduced if a computer file can be sent to the machining supplier and a three-dimensional view of the part is used instead of a two-dimensional view.

Design Errors. The sixth and last, but certainly not least, reason for failures is design errors. These errors include all engineering disciplines involved in the creation of a design:

- Design
- Materials
- Stress
- Manufacturing
- Quality

Since the 1970s, the U.S. government has exerted considerable pressure on car companies to reduce emissions from their automobiles. Many of the car companies redesigned vehicles, giving them diesel engines to reduce the emissions from a portion of their fleet and meet the total fleet emission standard set by the federal government, thus evading millions of dollars in fines. Unfortunately, this was a costly endeavor that required not only redesign but also retooling of the manufacturing line due to the differences between a diesel engine and a gas engine.

One car company decided to take a risk and not redesign the car—but simply replace the gas engine with a diesel one. They picked an unpopular model and figured that no one would buy the car or the diesel engine. Company engineers determined that that the diesels would start to have problems at about 80,500 km (50,000 miles). To the company's surprise and dismay, a gas shortage hit the United States and the diesels were completely sold. True to the engineering prediction, the cars started to come apart at about 80,500 km (50,000 miles) and the company had to replace each one.

Interestingly, the risk turned out to be cost effective. It took three to five years for the average consumer to reach 80,500 km (50,000 miles). By that time the company had saved millions by evading the federal fines. However, there was also some poetic justice. The car company finally decided to create a proper diesel engine design. Unfortunately, the combination of the poor reputation of their first diesel-powered car with the downturn in consumer interest proved to make the better designed diesel-powered car a losing model.

In the 1960s a metallurgical phenomenon was discovered when commercially pure (CP) titanium was accidentally welded to the titanium-6Al-4V alloy. The Ti-6Al-4V alloy represents approximately 70% of the titanium world market. Commercially pure titanium is probably another 20%. Even though you can weld them together, the dissimilar titanium material combination produces a buildup of titanium hydrides in the CP titanium via the rare process of uphill diffusion of hydrogen from Ti-6Al-4V to the CP titanium. The process occurs over time at room temperature, and the result is a layer of titanium hydrides. The titanium hydride precipitate is a very brittle, needlelike phase, and the precipitation mechanism is driven by the uphill diffusion mechanism. Due to this brittleness, the welded joints fail well below normal levels.

The phenomenon was noted on the Saturn S-IVB program when a Ti-6Al-4V propellant tank was pressure tested approximately 1.5 years after the qualification pressure test and failed catastrophically. The pressure test was at much lower pressure than the earlier qualification test. The tank had been accidently welded using CP titanium welding rods. Although this failure happened in the early 1960s, technical information on the subject is difficult to find. Manuals and handbooks indicate that you should be careful, but offer no specific warning or information.

One might imagine that the welding of dissimilar titanium alloys would have been banned from all designs. However, I have experienced this design on various programs over the last 25 years, including the Peacekeeper missile program, a Star Wars program, a commercial airliner program, and the ISS program. It was especially critical on the ISS program due to the 15-year lifetime. This is an example of design error where technical information exists, but people just do not know about it. Is it a lack of knowledge, poor engineering research, or simple apathy?

Another design error involves the Custom 455 CRES part described earlier. As noted previously, the physical cause of failure was the inadequate solution heat treat time at 830 °C (1525 °F). But the latent cause of failure was the design, which required the part to be solution heat treated.

Most Custom 455 stainless steel is purchased in the heat-treated condition, machined, and put into use. The only reason the part was being solution heat treated was because it had a hole drilled down the center to reduce its weight by 4 lb. The configuration of the hole, combined with the type of PH stainless steel and the lack of adequate inspection capability in a hole with a poor depth-to-diameter ratio, required that the part be solution heat treated and precipitation aged after rough machining. This led to the problem of poor mechanical properties.

The failure investigation and work to accept the poorly heat-treated parts was all generated by the desire to eliminate 4 lb. There are six of these parts on the ISS assembly that fits into the Space Shuttle cargo bay. The company most likely spent hundreds of thousands of dollars in order to cut 24 lb from an ISS assembly that probably weighed more than 30,000 lb. It would have been far simpler, much less costly, and technically better not to drill the hole.

Design choices, design errors. The technical expert explained the risks and the costs, and in the end it came down to saving 24 lb. Unfortunately, due to labor allocation practices, most designers are no longer involved with a program from concept to fabrication. Designers design something, then move on to a new program and are never around to see the first design built. Thus, they learn nothing about their designs, and someone else has to deal with subsequent problems. The result is that we see more design errors, many a duplicate of a previous one.

The design of any assembly generally requires fasteners. The final assembly of the ISS is performed in space by astronauts. Working in space presents new challenges and requirements to designers. For example, designers learned that cadmium plating cannot be used on fasteners in space. Although a very good plating for fastener lubrication and protection, cadmium will sublime in space, going straight from a solid to a gas.

The second choice was silver plating, also a very good material for fastener lubrication. Unfortunately, the atomic oxygen in the low earth orbit of the ISS oxidizes silver to a very fine, loose black powder that may deposit on thermal-sensitive areas of the ISS. Even worse, the astronauts could rub against the silver and bring the fine powder back inside the ISS, creating a safety hazard. Lastly, silver causes liquid metal embrittlement (LME) in titanium, and titanium fasteners were already in use on the ISS. Designers were again disappointed to discover that silver-plated fasteners had been removed from the list of approved fasteners.

The last choice for fastener lubrication in space was the application of molybdenum disulfide, a dry film lubricant. But this coating also presented the design team with issues. If molybdenum disulfide was left in the atomic oxygen of low earth orbit for too long, it became molybdenum trioxide, which is not a lubricating coating and indeed can be very abrasive. The team had to design protection for the fasteners with dry film lubricant so that when the astronaut used a fastener in orbit, it would work.

The interesting point is that none of designers was aware of these restrictions on various platings and coatings. Therefore, when determining the final assembly concepts, they were hampered in the selection of fasteners and lubrication.

Operational temperature presents another potential source of design error. As noted earlier, Custom 455 PH martensitic stainless steel was used on the ISS. This was an interesting choice, because the temperature fluctuation of the ISS as it orbits the earth every 90 minutes is −185 to +150 °C (−300 to +300 °F). The Custom 455 is acceptable for use at 150 °C (300 °F), but goes through a ductile-to-brittle transition in the range from −20 to −35 °C (0 to −30 °F). This is a common phenomenon for a martensitic stainless steel. So how was the Custom 455 selected for use? It was discovered that the stress engineers could not find any cryogenic tensile or fracture toughness data on Custom 455 at −185 °C (−300 °F), so they decided to use the cryogenic data for A286, a PH austenitic stainless steel that does not exhibit a ductile-to-brittle transition. This decision was completely in error and was a human cause of failure. As a result, many of the ISS components had to be reevaluated using reduced mechanical properties.

The last example of design error is a simple one. Machining drawings usually have a three-decimal dimension, such as 60.325 mm (2.375 in.) with a tolerance of ±0.254 mm (±0.010 in.). This is quite acceptable for a machined part, but not for castings, which have other desirable characteristics such as quick, inexpensive, and easy production and the capability to produce intricate shapes. Castings often are used in the as-cast condition, which requires a less precise dimensional callout of two decimals, such as 60.45 mm (2.38 in.), due to the casting process itself. Many designers forget this fact and put three-decimal dimensional callouts on the casting drawings. As a result, the castings cannot meet the drawing and the program loses time and money while the casting is submitted for approval through the discrepancy process. Once again, this is a human cause of failure.

There are hundreds of examples as to why failures happen. The ones presented here are just the tip of the iceberg. As a last thought, consider the real possibility that failures occur as a cascade of events, or failure modes, and that only the final event causes the final failure.

REFERENCE

1. G.H. Koch, M.P.H. Brongers, N.G. Thompson, Y.P. Virmani, and J.H. Payer, Direct Costs of Corrosion in the United States, *Corrosion: Fundamentals, Testing, and Protection,* Vol 13A, *ASM Handbook,* ASM International, 2003, p 959–967

Case History: Tacoma Narrows Bridge Collapse, Washington State (1940)

Rita Robison

The failure that brought an abrupt end to the search for "a slender ribbon bridge deck" also introduced the importance of wind dynamics to bridge design.

Background

Although the newspapers were filled with important stories from throughout the world on Saturday, November 8, 1940—stories such as the RAF bombing Berlin, the Luftwaffe bombing London, and Franklin D. Roosevelt returning to Washington D.C. for an unprecedented third presidential term—there was still room on the front page of the *New York Times* for a single-column story, "Big Tacoma Bridge Crashes 190 Feet into Puget Sound."

The Tacoma Narrows Bridge was the third largest suspension span bridge in the world and only five months old. The center span, measuring 853 m (2800 ft), stretched between two 130 m (425 ft) high towers, while the side spans were each 335 m (1100 ft) long. The suspension cables hung from the towers and were anchored 305 m (1000 ft) back towards the river banks. The designer, Leon Moisseiff, was one of the world's foremost bridge engineers. He and his partner Fred Lienhard had earlier developed the calculations for determining load and wind forces used by bridge designers everywhere.

Following the widespread design effort of the 1930s toward "streamlining" every product from teakettles to locomotives to airplanes, Moisseiff's intent was to produce a very slender deck span arching gently between the tall towers. His design combined the principles of cable suspension with a girder design of steel plate stiffeners—running along the side of the roadway—that had been streamlined to only 2.4 m (8 ft) deep.

The $6.4 million bridge had opened with much fanfare on July 1, celebrated as a defense measure to connect Seattle and Tacoma with the Puget Sound Navy Yard at Bremerton, Washington. Owned by the Washington State Toll Bridge Authority, the bridge had been financed by a Public Works Administration grant and a loan from the Reconstruction Finance Corporation, and had been constructed in only 19 months.

The bridge gained notoriety even before it opened, and was nicknamed "Galloping Gertie" by people who experienced its strange behavior. Forced to endure undulations that pitched and rolled the deck, workmen complained of seasickness. After the opening, it became a challenging sporting event for motorists to cross even during light winds, and complaints about seasickness became common.

Despite Moisseiff's reputation as a top-ranked engineering consultant, State and Toll Bridge Authority engineers were more than a little nervous about the behavior of the slender two-lane span, which was only 12 m (39 ft) wide. Its shallow depth in relation to the length of the span (0.9–853 m, or 3–2800 ft) resulted in a ratio of 1:350, nearly three times more flexible than the Golden Gate or George Washington bridges. Engineers tried several methods to stabilize, or dampen, the oscillations—the up-and-down waves of the deck.

The first method involved attaching heavy cables—called tie-down cables—from the girders and anchoring them with 45,900 kg (50 ton) concrete blocks on shore. The cables soon snapped, and another set installed in a second try lasted only until the early morning hours of November 7. A more successful method, a pair of inclined stay cables connecting the main suspension cables to the deck at mid-span, remained in place but proved ineffectual. Engineers also installed a dynamic damper, a mechanism consisting of a piston in a cylinder, that also proved futile as its seals were broken when the bridge was sandblasted prior to being painted.

Measurements and movie camera films taken over several months gave the engineers a good idea of how the bridge was moving in the wind. They charted the oscillations and vibrations, and discovered the movements were peculiar. Rather than damping off (dying out) very quickly as they did in the Golden Gate and George Washington bridges, Galloping Gertie's vibrations seemed almost continuous. The puzzle, however, was that only certain wind speeds would set off the vibration; yet there was no correlation between the vibration and the wind speeds.

Led by Frederick B. Farquharson, professor at the University of Washington's engineering school, the study team applied actual measurements of the bridge movements to a scale model, hoping to find ways to stabilize it. They suggested installing additional stabilizing cables, attaching curved wind deflectors, and drilling holes in the girders to permit wind to pass through. Unfortunately, the report was issued only a week before the bridge collapsed.

Interest in the phenomenon rose with the onset of the brisk fall winds pushing through the valleylike narrows that lie between the cities of Tacoma and Bremerton. The public as well as the engineers kept watch, and when the bridge finally snapped, it became one of history's most documented disasters. Cameras—still and movie—recorded the collapse on film. A newspaper man was a mid-span survivor, supplying a firsthand account of the disaster.

Details of the Collapse

Witnesses included Kenneth Arkin, chairman of the Toll Bridge Authority, and Professor Farquharson. As Arkin later recalled, that morning he had driven to the bridge at 7:30 to check the wind velocity, and by 10:00 saw that it had risen from thirty-eight to forty-two miles per hour while the deck rose and fell 0.9 m (3 ft) 38 times in one minute. He and Farquharson halted traffic, watching while the bridge, in addition to waving up and down, began to sway from side to side. Then it started twisting (Fig. 4).

Meanwhile, Leonard Coatsworth, the newspaperman, had abandoned his car in the middle of the bridge when he could drive no further because of the undulations. He turned back briefly, remembering that his daughter's pet dog was in the car, but was thrown to his hands and knees. Other reporters described him, "hands and knees bloody and bruised," as he crawled five hundred yards while the bridge pitched at forty-five degree angles and concrete chunks fell "like popcorn." The driver and passenger of a logging truck told a similar tale of having jumped to the deck and crawled to one of the towers, where they were helped by workmen.

By 10:30 the amplitude (distance from crest to valley) of the undulations was 7.6 m (25 ft) deep and suspender ropes began to tear, breaking the deck and hurling Coatsworth's car and the truck into the water. When the stiffening girder fell 58 m (190 ft) into Puget Sound, it splashed a plume of water 30 m (100 ft) into the air. Within a half hour, the rest of the deck fell section by section, until only the towers remained, leaning about 3.7 m (12 ft) toward each shore. Overlooking the bridge was an insurance company billboard that bragged, "As safe as the Narrows

Fig. 4 The center span of the Tacoma Narrows Bridge writhes under a gale-force wind. Moments later the bridge collapsed.

Bridge." The slogan was covered up before the end of the day. The only casualty of the collapse was the dog.

Moisseiff's first public comment was, "I'm completely at a loss to explain the collapse."

Charles E. Andrew, chief engineer in charge of construction, said that the collapse was "probably due to the fact that flat, solid girders were used along the sides of the span." He wanted to make clear that his original plans called for open girders, but "another engineer changed them." He compared Galloping Gertie with the New York's Whitestone Bridge, which had been completed a year earlier. It was the only other large bridge designed with web-girder stiffening trusses, "and these caused the bridge to flutter, more or less as a leaf does, in the wind. That set up a vibration that built up until failure occurred."

The Whitestone, now known as the Bronx-Whitestone, was indeed very similar to the Tacoma Narrows. However, the stiffening trusses were twice as heavy and the deck twice as wide. Its designer, Othmar Ammann, had consulted Moisseiff, as had the designers of such structures as the Golden Gate and the San Francisco-Oakland Bay bridges. The Whitestone is 1220 m (4000 ft) long, with a 701 m (2300 ft) main span and 358 m (1175 ft) high towers. It, too, had early oscillation problems, but they had been successfully damped with a device similar to the one tried on Galloping Gertie.

After the Tacoma failure, Ammann realized the danger of the too-slender deck, and insisted that a new steel truss be superimposed onto the deck. His redesign, which also added cable stays between the towers and deck to prevent twisting, stabilized the bridge. In 1990, a tuned mass damper (a fairly recent invention) was added to the deck during a rehabilitation.

The first investigations into the collapse of Tacoma Narrows detailed how the bridge had come apart. For months the motions, while disturbing, had been symmetrical and the roadway had remained flat (Fig. 5). The lampposts on the sidewalks remained in the vertical plane of the suspension cables even as they rose, fell, and twisted. But on November 7, a cable band slipped out of place at midspan, and the motions became asymmetrical, like an airplane banking in different directions. The twisting caused metal fatigue, and the hangers broke like paper clips that have been bent too often.

Impact

But what caused Galloping Gertie to twist so violently when other bridges had survived gale-force winds? Engineers looking into the problem of the twisting bridges were able to explain that winds do not hit the bridge at the same angle, with the same intensity, all the time. For instance, wind coming from below lifts one edge, pushing down the opposite. The deck, trying to straighten itself, twists back. Repeated twists grow in amplitude, causing the bridge to oscillate in different directions.

The study of wind behavior grew into an entire engineering discipline called aerodynamics, parallel to that of the airplane industry. *Vortex shedding* and *flutter* were added to the vocabulary. A vortex is a spiral that can be seen in the wake of a ship or, in a wind tunnel, by wisps of smoke added for the purpose. Some vortices do not affect oscillation, but others form a flutterlike pattern that has the same frequency as the oscillating bridge. Eventually no bridge, building, or other exposed structure was designed without testing a model in a wind tunnel. With the development of graphic capabilities, some of this testing is now done on computers.

Dozens of papers are published each year about these subjects; nevertheless, misconceptions continue. In 1990, K. Yusuf Billah of Princeton University noticed that currently popular physics textbooks were using an incorrect version of the cause of the Tacoma Narrows collapse. He enlisted Robert H. Scanlan of Johns Hopkins University, and they "set the record straight" in the February 1991 issue of the *American Journal of Physics*. The textbooks were claiming "forced resonance (periodic natural vortex shedding)" as the cause. Instead, Billah and Scanlan pointed out, "the aerodynamically induced condition of self-excitation [vibration] was an interactive one, fundamentally different from forced resonance."

Fig. 5 For months the motions, while disturbing, had been symmetrical and the roadway had remained flat. Such twisting motions caused metal fatigue, resulting in the collapse on November 7, 1942.

Two years after the collapse, the remains of the Tacoma Narrows Bridge were scrapped. In 1950, the state opened a new $18 million bridge designed by Charles Andrew and tested in wind tunnels by Farquharson and other engineers (Fig. 6). Its four-lane, 18 m (60 ft) wide deck and stiffening trusses 7.6 m (25 ft) deep form a box design that resists torsional forces. Excitation is controlled by hydraulic dampers at the towers and at midspan.

The Tacoma Narrows Bridge collapse remains one of the most spectacular failures in the history of engineering. It is certainly one of the best known because of the movie film and widely reproduced still photos, but as Billah and Scanlan demonstrated, the aerodynamic details tend to become blurred. The new science of wind engineering, however, is now routinely applied to every type of structure.

Fig. 6 This new bridge was opened in 1950. Its four-lane deck and stiffening trusses form a box design that resists torsional forces. Excitation is controlled by hydraulic dampers at the towers and at midspan.

SELECTED REFERENCES

- D. Alexander, "A Lesson Well Learnt," *Construction Today,* Nov 1990, p 46
- C.H. Ammann, T. von Karman, and G.B. Woodruff, "The Failure of the Tacoma Narrows Bridge," Report to the Federal Works Agency, March 28, 1941
- Billah, K. Yusuf, and R.H. Scanlan, "Resonance, Tacoma Narrows Bridge Failure, and Undergraduate Physics Textbooks," *American Journal of Physics,* Feb 1991, p 118–124
- F.B. Farquharson, "Collapse of the Tacoma Narrows Bridge," *Scientific Monthly,* Dec 1940, p 574–578
- R.R. Goller, "Legacy of Galloping Gertie 25 Years After," *Civil Engineering,* Oct 1965, p 50–53
- D.C. Jackson, *Great American Bridges and Dams,* Preservation Press, 1988, p 327–378
- "Big Tacoma Bridge Crashes 190 Feet into Puget Sound," *New York Times,* Nov 8, 1940, p 1
- I. Peterson, "Rock and Roll Bridge," *Science News,* 1990, p 244–246
- H. Petroski, "Still Twisting," *American Scientist,* Sept–Oct 1991, p 398–401

Case History Discussion

As late as 1991 engineers were still debating the 1940 failure of the Tacoma Narrows Bridge in Puget Sound. As technology advances, failure causes are understood more completely through the use of better analytical tools and better examination tools. Years later we realize that some conclusions reached in the past were wrong. Sometimes the past teaches us what we need to know. Sometimes the past wasn't smart enough to know what was going on. That's what happened here.

This spectacular bridge failure resulted in far-reaching ramifications. The introduction of the concept of self-harmonics was one ramification. The concept of self-excitation came later. Another ramification was the importance of aerodynamics on stationary objects like bridges, which had not been considered before this bridge failure. Because of this failure, wind-tunnel testing would be applied to all structures, not just bridges.

The bridge was seen to "gallop" many times before the failure, even during construction. Everyone knew this was happening, but no one thought it was important. The engineers certainly considered the bridge safe enough to drive out and do measurements of the movement.

The physical cause of the bridge failure was metal fatigue. The human cause of the failure was the slender bridge design with the type of railing used. The latent cause of the failure was the "streamlining" design movement of the 1930s and its application to this bridge, which spanned a windy gorge. The problem disappeared when design changes were made including a nonsolid railing either slatted or gapped sides. The original bridge design created lift, and the lift created a cable stretching problem, which created a harmonic excitation—that old cascading failure concept again.

CHAPTER 3

Aspects of a Failure Investigation

IMPORTANT ASPECTS of failure investigation addressed in this chapter include:

- Benefiting from the use of statistics
- Understanding databases and data analysis techniques
- Determining the root causes of failures
- Deciding when a failure investigation should be performed
- Understanding and using problem-solving techniques
- Planning a failure investigation

Relevant Statistics

Statistics are an important part of a failure investigation. It is in the best interest of a company to keep statistics on failures—including type of failure, material, root cause, and so forth—in a searchable database.

Table 1 presents statistics on failure causes in some industry investigations. According to the table, the leading cause of failures is improper material selection. Things certainly have changed since 1948, when material selection was not even thought of as a potential failure root cause.

Table 1 Causes of failures in some industry investigations

Failure origin	Percentage
Improper material selection	37
Fabrication defects	16
Improper heat treatment	14
Design error	12
Unexpected operating conditions	9
Inadequate environmental control	5
Improper inspection/quality control	4
Wrong material	3

But what exactly does "some industry investigations" mean? A person or company cannot tell from the table title whether the data apply to their situation.

The various failure origins listed in Table 1, including improper material selection, fabrication defects, design error, and so on, are all definitive causes of failures. The data are well presented, the numbers appear to make sense, and the percentages add up to 100. But the data are not acceptable for use by anyone except the person who developed the table.

Table 2 narrows down the database to aircraft components. This table is more specific about the sample population from which the statistics are derived. A review of the data indicates that improper maintenance is now the number one failure cause; improper material selection does not even make the list, except perhaps in the category of "undetermined cause." But what is the significance of the term "laboratory data" in the table title? A person cannot tell.

Table 2 also has well-described failure origins, and the data are again well presented and appear to make sense. One interesting origin in this table is labeled "abnormal service damage." It is difficult to understand what this category means without knowing what constitutes normal service damage. "Undetermined cause," on the other hand, is a useful category for statistical data, because sometimes a failure investigation does not reach a decisive conclusion.

Since each industry, and even each company, is different, what is pertinent and relevant to one is meaningless to another. Each needs to keep its own statistics and draw upon outside sources as applicable. A company's own failures are usually its best source of data.

Statistical data must be relevant before anyone will pay attention to it. For example, suppose a medical statistic states: "Men in their 40s have the most fatal heart attacks." When women read this statistic, they do not pay attention after the first word. When men who are older than 49 or younger than 40 read this statistic, they don't pay attention after the fourth word. Only men in their 40s pay attention to the entire statistic. Data have to be relevant or the audience will tune out.

Table 2 Causes of aircraft component failures (laboratory data)

Failure origin	Percentage
Improper maintenance	43
Fabrication defects	18
Design deficiencies	15
Abnormal service damage	11
Defective material	8
Undetermined cause	5

Creating a Database

The most difficult part of creating a database is to begin. Although statistics are great to have as a resource, collecting failure investigation data and maintaining a database are rarely given a high priority by management. This mindset needs to change.

Collecting the Data. First, start within your own company and create a database of failures. Organize the database for particular needs and trends. The database will indicate which areas require more attention. Second, expand into the industry in which your company participates. Many industry failures are a matter of public record. In addition, failures can be found documented in conference proceedings and published papers. Incorporate only relevant data. A good idea is to make sure the source of each data point is noted. Data are only as good as their pedigree. Lastly, search the Internet for data. However, always remember that such information may not have been checked or validated and can be incorrect. If not from a known and trusted source, Internet data should be verified from multiple sites or from a book, or by contacting the company.

Let's look at Table 3. Similar to Table 1, it presents a set of statistics on the frequency of failures in some industry investigations. But these data are based on the type of failure mechanism instead of the failure origin. Was this an attempt to separate the physical root causes from the human and latent root causes? Maybe, but since it is not our data, we cannot be sure.

The data in Table 3, which are once again well presented, indicate that the top 54% of failure causes are life issues of corrosion and fatigue. However, two other kinds of corrosion also are noted. So what does the top category of corrosion mean? And again, to what industry does the table refer? Are the data applicable to your industry? Good table, but it generates more questions than answers.

Table 4 is another set of statistics on failures of aircraft components, similar to Table 2. But this time the data are based on the type of failure mechanism instead of the failure origin. Was this another attempt to separate the physical root causes from the human and latent root causes? Once again, we cannot be sure.

Table 3 Causes of failures in some industry investigations

Failure mechanism	Percentage
Corrosion	30
Fatigue	24
Brittle fracture	17
Overload	10
High-temperature corrosion	8
Stress corrosion/corrosion fatigue/hydrogen embrittlement	5
Creep	4
Wear, abrasion, and erosion	2

The data in Table 4 show that the top 77% of the failure causes are life issues of fatigue and overload. However, these data do not indicate if the failure was acceptable according to the component's design criteria. For life issues, it is important to know if the failed part achieved the acceptable life criteria. When a company designs a part that operates in a fatigue environment, the part is expected to be replaced after so many cycles, and to fail after additional cycles. If a component fails by fatigue, it's possible that someone didn't replace a piece of equipment per the maintenance time schedule and that the component did not fail prematurely. The root cause of the failure becomes a human cause instead of a physical cause.

Using the Data. The importance of a failure analysis database and the corresponding statistics is to provide direction. The statistics can point to problem areas in your company. If you can take all the failures in the database and present them on a statistical chart like a Pareto diagram (Fig. 1), the data may indicate that 80% of the company failures revolve around the top three failure mechanisms. From an economic point of view, the statistical database has just provided the direction as to how the company can improve profitability by indicating where the company should spend money and labor. No company can solve every problem, so it is prudent to attack the most technically detrimental or costly problems first. Statistics can provide this direction.

Including the cost associated with failures will make the statistical database even more beneficial. This information may affect decisions regarding money and labor. In Table 4, for example, fatigue is the failure cause that requires the most attention based on frequency of occurrence. However, if an additional table column indicated that fatigue cost the company $5 million a year due to maintenance of aircraft components reaching the end of life, but stress rupture cost the company $20 million a year due to catastrophic engine failure and aircraft loss, then the company focus might be on the stress rupture failure cause, even though it occurs only 2% of the time.

Statistics can also point to trends over time or over the life cycle of a certain product line. Tracking the life of a particular product design using the number of cycles or years of service of each failure may provide information as to its expected life. In addition, tracking a product's manufacturing date may provide a correlation of failures to the time of year the product was manufactured. For example, many companies in the alumi-

Table 4 Causes of aircraft component failures

Failure mechanism	Percentage
Fatigue	60
Overload	17
Stress corrosion	7
Excessive wear	6
Corrosion	5
High-temperature oxidation	3
Stress rupture	2

num industry indicate that their success in casting a certain type of alloy depends on the time of year. Specifically, they cast the 7050 alloy more successfully during the winter months when the humidity is lower, because pouring during high humidity can lead to a hydrogen issue. This appears to be an item of discussion in the aluminum industry, but there is little published information.

Comparing company statistics to industrial data will also indicate if a problem is company specific or industrywide. You can benchmark your company against the industry.

Presentation of Statistics. Numbers can be misleading, so it is important that data be presented carefully. Sometimes it is best to request the presentation of information in absolute values instead of percentages. For example, a sales engineer may provide the statistic that the company increased its yearly number of new products by 100%. That sounds wonderful until you realize that the absolute increase was the introduction of two new products versus last year's one new product, which is not so impressive. Make sure the statistics make sense; do not follow them blindly.

Statistical methods must also be selected with care. Skillful "design of experiments" is essential. Engineers frequently must decide whether it is necessary to perform a full factorial set of tests or whether it's accept-

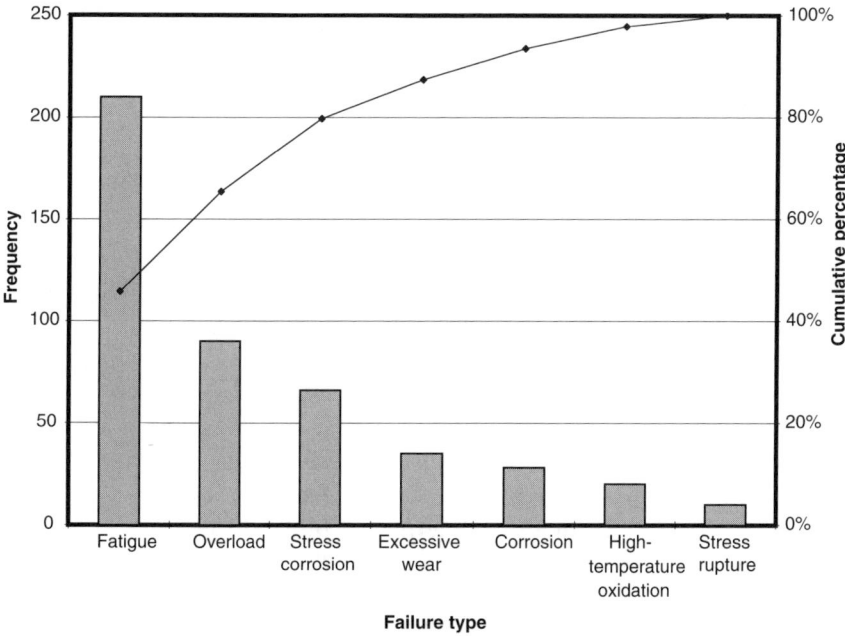

Fig. 1 A Pareto diagram is used to plot the relative importance of the differences between groups of data—in this case, the frequency of different types of failure at a company. The bars show the number of counts for each failure type, and the data points show the cumulative percentage (from left to right). In this example, the three most frequent failure types at the company account for 80% of total failures.

able to reduce the test matrix and perform a statistical analysis of the data. A full factorial set of tests can be expensive and time consuming, but will provide an absolute answer to the question of which variable most affects the outcome. Statistical methods such as Taguchi analysis are designed to provide a trend determination and ranking of variables. Statistic-based testing is less expensive and time consuming than a full factorial test program; however, the result is not a final answer. In addition, be aware that a statistical analysis method can become "popular" and used for problems to which it does not apply. Lastly, a full factorial test program may provide information to solve a future problem. The test matrix breadth may be enough to encompass the new problem.

Sometimes the best answer is a combination of the two approaches. If you have many variables, use Taguchi analysis to determine the trend analysis and variable ranking. Then take the top three or four variables and perform a full factorial test program to determine the final answer. Many excellent statistical tools are available in the technical area of design of experiments. It's best to speak to an expert and research which tool best fits your problem.

Database Information. What information should be in your company's statistical database? How many fields should the database have? The key is that the database must be pertinent to the company's needs. Ask potential database users—management, sales, and quality personnel, for example—to indicate what information they would search. This will help to determine which fields should be included and which statistical methods are required for data analysis.

For example:

Component (part number, serial number)	P/N 8BB3445, S/N 12345
Manufacturing date	1/02/2000
Date of failure	2/23/2002
Material (heat treatment, heat number)	321 CRES, annealed
Where (what plant, city, state, country?)	Cleveland, Ohio
Time of year	Winter
Type of failure (service, maintenance, testing)	Service
Failure mechanism	Overload
Submechanism	Ductile
Data source (company, industry, Internet)	Internal
Cascading failure (yes/no)	No
Achieve design life (yes/no)	No
Root cause (physical, human, latent)	Human: wrong load applied

The types of failure mechanisms to track in your statistical database can be overwhelming. The mechanisms change names and become further subdivided each passing year as more technical information is acquired. Here is a partial listing:

- Ductile and brittle fracture
- Fatigue: high cycle, low cycle, thermal, corrosion, etc.
- Corrosion: uniform, pitting, selective leaching, intergranular, crevice (O_2 starvation), galvanic, concentration cell, temperature differential, bacterial and biofouling, erosion affected, etc.
- Liquid erosion: cavitation, impingement, melting, etc.
- Distortion (plastic or elastic)
- Stress corrosion: stress, environment, material susceptibility
- Liquid metal embrittlement
- Solid metal induced embrittlement
- Elevated temperature: creep, stress rupture, fatigue, creep-fatigue, etc.
- Hydrogen damage: embrittlement, blistering, internal hydrogen precipitation (flakes), hydride formation, etc.
- Radiation
- Combinations of various failure mechanisms

The important point to remember is to make your database usable for company-related failures. Creating a database that cannot be used is a waste of time.

Why Is a Failure Investigation Performed?

Many young engineers new to the world of failure investigation answer this question very directly: "My boss told me to do it." That answer is correct, of course, but the true purpose of a failure investigation is to determine the root cause(s). Determination of the root cause is good engineering practice that crosses functional boundaries within a company and is an integral part of the quality assurance and continuous improvement programs. Note that one or many root causes may underlie a single failure. Many failure investigations search out the "smoking gun," the one root cause, but few are that simple.

A proper failure investigation is not a "science project." There also is a distinct difference between a failure investigation and a metallurgical evaluation. A metallurgical evaluation is always part of a failure investigation, but it determines answers only to the metallurgical questions posed in the investigation and most times leads only to the physical root cause. A complete failure investigation determines the root cause of the failure—physical, human, or latent—and recommends appropriate corrective actions.

Why Determine Root Cause?

The most obvious reason for discovering the root cause of a failure is to determine the fault or innocence of a company or person during litigation (although litigation sometimes determines who *can* be blamed rather than who *is* to blame). Often a company is proving its own innocence. Once fault is established, monetary assessment usually follows.

However, a more common reason for discovering the root cause is so that corrective action can be implemented to prevent future occurrences, thus saving a company time and money. It is also possible to determine whether the failure was unique or symptomatic of a widespread, systemic problem.

Chapter 2 presented an example of a fabrication failure that involved a part in the hot gas portion of the Space Shuttle main engine high-pressure oxidizer turbopump. Made from 321 CRES tube in the shape of a ring with multiple small cooling jets welded into it, the part exhibited a series of cracks. Nevertheless, the Space Shuttle flight was launched on schedule for two reasons. One, the physical cause was presented to NASA and accepted. And two, because the specific physical cause for the failure had been identified, and examination of other turbopumps revealed no cracks in similar components, the problem was not deemed systemic to the fleet.

Another example concerns flaws in large 7050 aluminum ring-rolled forgings used on expendable launch vehicles. The 7050 aluminum alloy is used because it can be fabricated in thick cross sections and still be heat treated to an adequate strength. The part in question requires the pour of an 8165 kg (18,000 lb) ingot to make a 1361 kg (3000 lb) forging to machine to a 156 kg (300 lb) part. An incredible amount of material is lost during manufacturing. Because the 1361 kg (3000 lb) forging cross section is so large, ultrasonic inspection for internal flaws is limited to a Class A level, or a maximum singular flaw size of approximately 2.0 mm (0.08 in.). The finished machine part is then dye penetrant inspected. It was during this surface inspection that flaws were discovered. If the parts could not be repaired, the company would experience a schedule loss of 8 to 12 months (not to mention the loss of $200,000 in forging and machining costs).

The failure investigation determined that the flaws were created during either the ingot fabrication or forging process or both, but that the 7050 material was not a factor. As part of the investigation, other large ring-rolled forgings made from other aluminum alloys, such as 7075 or 2219, were evaluated as to percentage of defects discovered during dye penetrant inspection of the machined detail part. The intent was to determine if there was a systemic problem or one just confined to the large 7050 ring-rolled forgings. Dye penetrant inspection found flaws in all the aluminum alloys, but the 7050 alloy had the highest percentage of occurrence. The company could find nothing in the literature to indicate why this might happen. It was during this investigation that the aluminum companies shared the concern noted earlier about pouring 7050 in the wintertime. So, in the end it was a widespread problem in the industry, but unique to the one 7050 alloy—an interesting discovery and conclusion.

When Is a Failure Investigation Performed?

The answer to this question depends on two aspects of any particular failure.

Relative Failure Importance. Table 5 provides examples of different levels of importance and the resultant decisions. The failures listed are not very important. No one thinks twice about throwing away a pencil if it breaks during use. In our electronic world, how many working people actually use pencils? A broken pencil is more important to school-age children and young adults, where usage is higher and supplies more limited. (I remember breaking my pencil in half on purpose to share with a classmate who did not have one.)

Breaking an ordinary pen also ranks low in importance, though there may be more important collateral damage to clothing. Personally, I have learned never to carry a pen in my shirt pocket anymore. However, breaking the pen your great-grandmother used in 1903 is a different matter. There is little doubt that you will attempt to fix the pen due to its sentimental importance. You may even call on an expert across the country, or perhaps in Europe. Time and money are not an issue.

Breaking a scuba knife is an interesting example. Scuba knives are usually worn by divers for emergencies such as freeing oneself if entangled in kelp or discarded fishing lines. If the knife breaks, you probably will at least complain to the manufacturer, because your life depends on this equipment. In wreck diving, the knife may also be used to free a diver from loose wreckage. In this case, the knife's importance changes with geographic location. Wreck diving on the West Coast is done in relatively shallow waters, under calm conditions, and on only a few wrecks, many of which have been made safe for divers. These divers depend on the scuba knife as their primary tool. Wreck diving on the East Coast, however, is done in deeper, rougher waters and on wrecks dating back hun-

Table 5 Examples of relative failure importance

Failure	Importance	Resultant decision
Break a pencil	Low	Throw pencil away, get a new one; more important to school-age children and young adults
Break a pen	Low	Throw pen away, buy a new shirt, complain to your coworkers, never use that kind of pen again
Break the pen your great-grandmother used in 1903	High	Determine problem and get it fixed; money not an issue
Break a scuba knife during a dive	High	Complain to the manufacturer and ask for a free replacement, or buy a more reliable knife; different importance for East Coast vs. West Coast divers

dreds of years. East Coast wreck divers carry a large tool bag containing various implements for emergencies. These divers would consider a broken scuba knife to be of low importance because they have an entire array of better tools more properly suited for the emergency at hand.

Now review the failures in Table 6. These failures are more important than the ones described in Table 5.

Many people keep their automobiles for many years. In the Rust Belt, wear and tear on the vehicle's body claim many cars with relatively low mileage. In states with milder climates, like California, a car may accumulate many miles along the way. If a car engine were to blow up after 150,000 miles, this would be important to most people because of the importance of their car to their life, the cost of replacing the engine or the car, and the nuisance factor. Even so, most people would feel that the 150,000-mile engine lifespan was acceptable. They would check to see if they had purchased the extended warranty, and then move on to the decision of getting the engine repaired or purchasing a new car.

What if a car engine blows up at 500 miles? The importance to the owner is the same for the same reasons noted previously. However, no one would deem a 500-mile lifespan acceptable. The owner would expect, perhaps demand, a new engine if not a new car, depending on the state's "lemon laws." The owner would complain that the car was junk, the car dealer a fraud, and the car company guilty of endangering the public. The point to this example is that the importance of a failure, or a failure investigation, is tied to the expected lifespan of the product in question.

The last example in Table 6 is a tragic one: the loss of an aircraft to the ocean. The public's expectation is that flying is safer than driving a car. Therefore, an aircraft crash, and the associated failure investigation, is of very high importance. The 1996 loss of TWA Flight 800, carrying 230 people from New York to Paris, was concluded to have been caused by a stray wire in the center wing fuel tank that ignited the flammable fuel/air mixture. The official failure investigation performed by the National Transportation Safety Board required more than four years to complete. However, many unofficial investigations that believe TWA 800 was shot down by a missile are still active in late 2003. When was the last time a fatal car failure investigation required four years?

Table 6 Additional examples of relative failure importance

Failure	Importance	Resultant decision
Car engine blows up after 150,000 miles	High	Check extended warranty; replace engine or buy new car
Car engine blows up after 500 miles	High	Demand new engine or new car from manufacturer, complain like nobody's business
Airplane crashes into the ocean	Very high	Suffer loss of plane and personnel, reputation, future business; deal with lawsuits; answer to government agencies

The TWA 800 accident also provides a practical example of a statistical database put into action. The importance of that failure investigation as well as other airline crash investigations led to the creation of an Internet database. In May 2001, Air Safety Online released a groundbreaking study on U.S. airline safety and launched a new aviation accident database (Ref 1). Another Web site tracks helicopter incidents (Ref 2).

Company involvement constitutes the second reason why a failure investigation is performed. If an investigation has the backing and interest of the company CEO, then getting laboratory priority, financial support, and sufficient help will not be a problem. However, the CEO usually gets involved only with high-profile cases that are important to the company. Then again, you may end up doing a failure analysis on the 1903-era pen that once belonged to his great-grandmother. A failure investigation on test lab tooling will not find such a champion.

Benefits of a Failure Investigation

An investigation that discovers the root cause or causes of a failure is an important and essential part of a quality program such as Total Quality Management and Continuous Improvement. Other benefits of a failure investigation include assisting in a redesign, solving a manufacturing problem, saving money, saving time, and, in some cases, saving lives. On occasion I've found myself involved with a team working on a problem where a redesign is under way before the failure investigation is complete. The designers have no idea what the root cause is but have come up with 54 ways to correct it.

Puzzles and Problem Solving

By nature, engineers are problem solvers. It is what separates us from other people. To demonstrate a point, consider the five puzzles shown in Fig. 2 to 6.

Imagine that each puzzle has been given to you to solve. The puzzle is in a brown envelope so you cannot see how difficult it is, and you are given 60 seconds to solve it. The goal is to complete the puzzle in the time allotted.

What do the puzzles tell us? Consider that each is like an individual failure investigation.

The puzzle in Fig. 2 represents the perfect failure investigation. All the correct pieces are present, they all fit together well, and there is enough time and labor available to complete the investigation on time. It is very nice. Please take a good look at this failure investigation, because it does not exist in real life. There is no failure investigation where all the pieces are there, they all fit together, and the answer just stares you right in the face.

The puzzle in Fig. 3 has one piece missing. This puzzle represents a failure investigation where some of the pertinent data or information is missing or was not discovered due to time or labor limitations. You spend time and effort checking the bag and the floor under your chair in order to make sure the piece is really missing. Is there enough information to solve the riddle of the puzzle? Maybe so, maybe not.

The puzzle in Fig. 4 has a few extra pieces. This puzzle represents a failure investigation where too much data or information was discovered or provided—information collected along the way that the engineer thought applied to the failure or that someone told the engineer was pertinent. You have to consider it all, looking at it and evaluating it before weeding it out and determining that an extra piece is not part of the puzzle or failure investigation. You cannot arbitrarily throw it away, because it might be relevant. In addition, during a review of the failure analysis

Fig. 2 The perfect failure investigation

someone may ask about the piece of information and you must account for it and indicate why you chose to disregard or use it. During the investigation, the extra information wastes time and effort. Having the extra piece is just as confusing as having one piece too few.

The puzzle in Fig. 5 has one black piece. The piece fits the shape of the puzzle, but the information it contains is not correct. This puzzle represents a failure investigation where incorrect or misleading information is discovered or provided. The black piece represents corrupted data, wrong data, bad data, or perhaps false or purposefully misleading data. Perhaps one of the eyewitnesses is lying or trying to disguise the true nature of the failure. Perhaps certain test data are wrong because the test was performed incorrectly. Can you tell when information is misleading

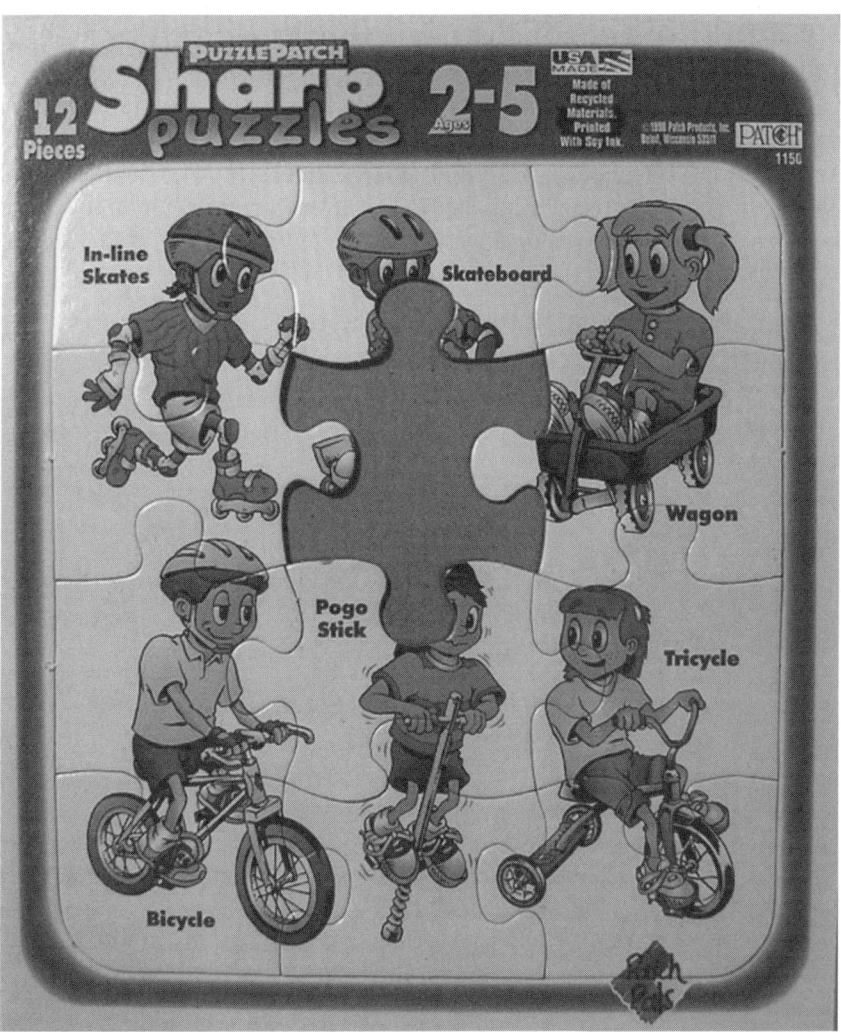

Fig. 3 Missing information

or incorrect? Sometimes. Does it stop you from solving the puzzle or the failure investigation? Sometimes. This puzzle teaches us that you can get information that appears to be right, appears to fit, but is wrong—and even though you can finish the failure investigation, you cannot get the right answer.

The last puzzle (Fig. 6) is most analogous to a real failure investigation. There are actually two puzzles in the envelope. It is not unusual to undertake an investigation where what the customer asks you to solve is really not the problem that you end up solving. Many times the customer does not really know what failure investigation they want to perform, or what question they are looking to answer. Being told to solve the puzzle does not provide sufficient information, since the envelope contains two puzzles. Which one are you to solve? In addition, there is not enough information to solve either puzzle. You could concentrate on both and spend double the time and effort, but in the end you are wasting time and effort, because you did not receive enough direction.

Or perhaps the customer could not provide adequate direction and you have all this extra stuff, which may not be bad information but has nothing to do with solving the puzzle and simply mucks up the situation. The puzzle in Fig. 6 is a true failure analysis: The customer doesn't know what really needs to be solved or how to solve it and has you working on two

Fig. 4 Too much information

failures at once. So you have to figure out what is the right way to go, and you have all this extra stuff that may or may not be helping. This is closer to reality than anything else.

The time allocation of sixty seconds and the labor estimate of one engineer are not realistic for the Fig. 6 puzzle. The puzzle, like a real failure investigation, is missing pieces such as pertinent data, has extra pieces such as nonessential information, and has pieces that appear to fit but in reality are just misleading. Can you solve this puzzle? Yes. However, the level of confidence you have in the answer is lower than for the other

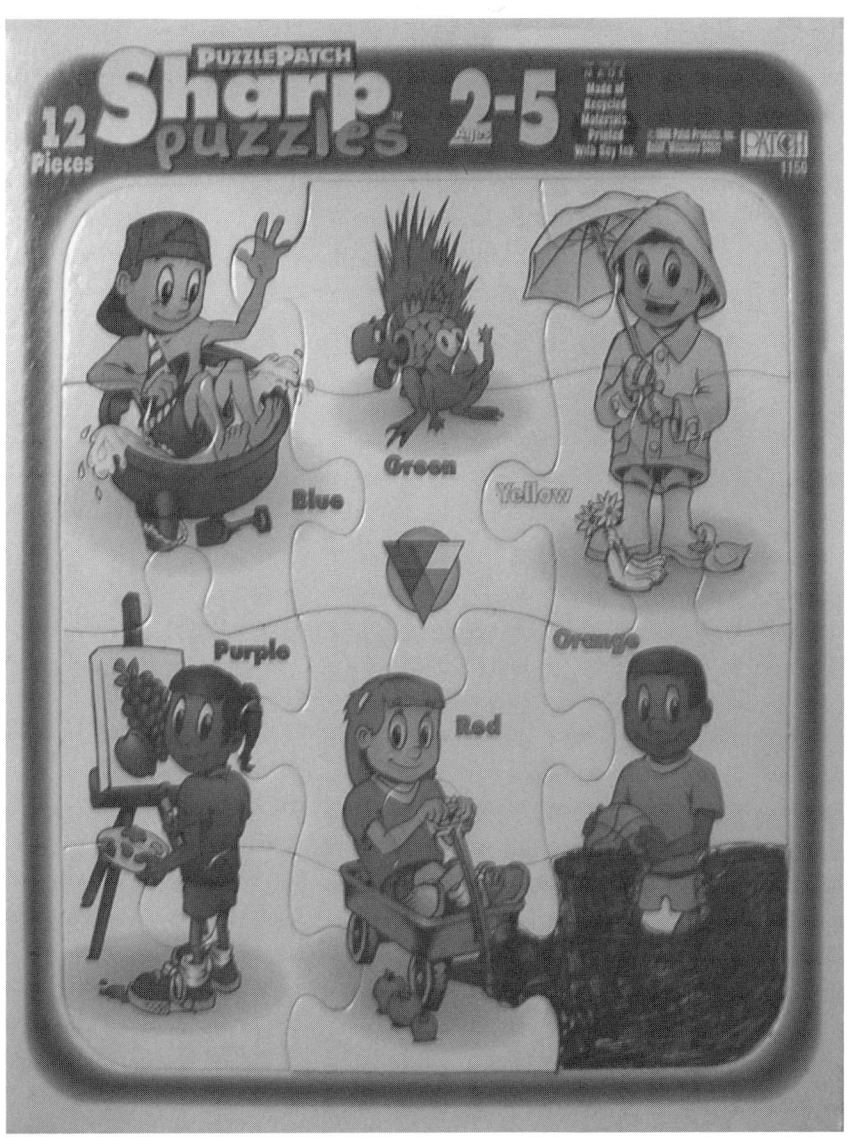

Fig. 5 Misleading or incorrect information

puzzles because of the reduced certainty that the puzzle can be solved, and because of the amount of confusing or misleading information presented.

The Four-Step Problem-Solving Process

The four-step problem-solving process that most engineers are taught is a good overall view of a failure investigation:

1. What is the problem?
2. What is the root cause of the problem?
3. What are the potential solutions?
4. What is the best solution?

All engineers learn this process, a sequence of events that you go through in your head whether you realize it or not. Engineers use this process to work out problems at the workplace and at home. In fact, some people think engineers are born with this problem-solving gene.

The first step of the four-step problem-solving process as applied to a failure investigation is to define the problem the investigation is to solve. This is perhaps the most important and difficult step. It will set the direction of the failure analysis, determine the time required, and determine the labor required. It is also the most difficult step because many engineers

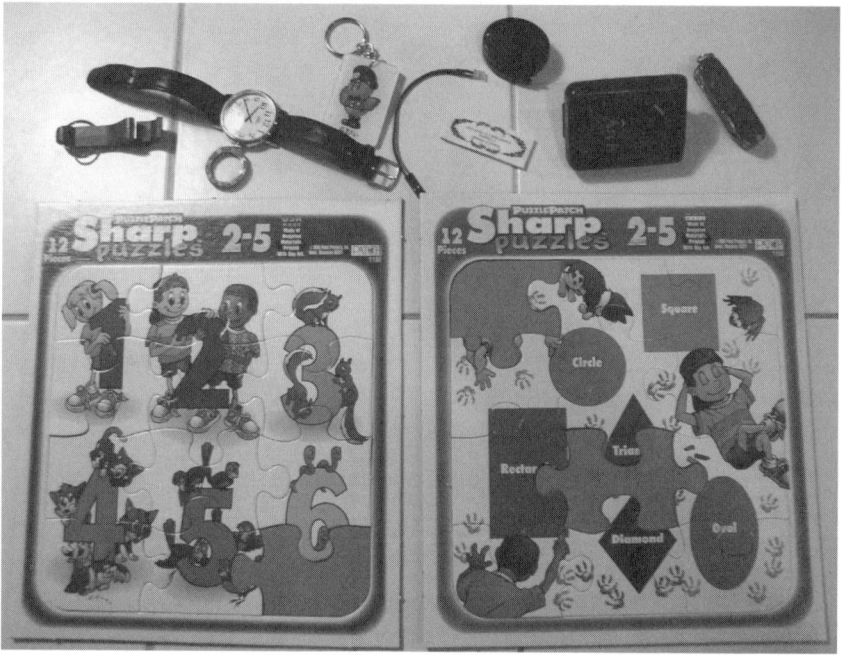

Fig. 6 The real failure investigation

launch into the technical evaluations they know how to do, before even thinking about what they are doing.

The second step is to define the root cause of the problem. This is the goal of any failure investigation. It is the step where all the hard work and planning occur. Many organizational and technical tools are at the disposal of a failure investigation, including brainstorming, fault trees, and testing. In many cases, it is also the step where the failure investigation damages samples, makes mistakes, and generally becomes lost. Pitfalls include poor organization, jumping to conclusions, and "silver bullet" theories from individuals who claim to know exactly why the failure occurred and who drive the investigation down a single path.

The third step can be accomplished only if steps 1 and 2 have been completed. Potential solutions may include:

- Design upgrades (make the product or system idiot-proof)
- Requirements relaxation
- Training
- Additional testing or relaxation
- Cautions or warnings
- Special operational or process actions

Many failure investigations never determine a problem's root cause. At that point, it becomes impossible to determine potential solutions. Yet many engineers try.

The fourth and last step is to recommend the best solution. The definition of "best" is determined by the requirements agreed to by the customer and the engineer performing the failure investigation prior to its start. Sometimes there is one definitive solution, and sometimes many solutions are employed.

A failure investigation is performed to (a) clearly identify and/or determine the failure's root cause(s) and (b) recommend corrective action(s). Many investigations end after step 2 of the four-step problem-solving process, after determining the root cause of the failure. However, it is just as important to undertake steps 3 and 4, identifying potential solutions and the best solution.

It is also a good idea to evaluate the best solution. Many failure investigations discover the root cause(s) and provide a recommendation for corrective action(s), but no one ever follows up to see if the corrective action worked. Chapter 4 will discuss how to evaluate the definitive recommendation, establish the implementation, and then develop a review process to determine that the recommendation actually worked.

Why Plan a Failure Investigation?

People spend a good part of their lives planning. Though they may not realize it, most people are planners. We plan vacation travel and business trips. We plan recreational activities such as scuba diving, parachuting, climbing, backpacking, and bicycle trips. How long am I going to be gone? What equipment do I need? Have I checked it out? What am I going to do each day? We plan home repair projects such as fixing a faucet or a door hinge, pouring cement, laying brick, and yardwork. Do I have all the correct tools? Do I need supplies? Do I need help? Should I get a professional to do this job?

In business, engineers plan buildings and bridges, highways and trains, cars and airplanes, even production of the smallest widget. Engineers plan the design, material delivery, manufacturing, and work schedules. They plan their entire day. There are three basic groups for the projects on an engineer's desk at any one time:

- Things that *will* get done today
- Things that *might* get done today
- Things that *will never* get done today

Engineers plan everything, except failure investigations. But that is about to change.

Timeline of a Failure Analysis

All failure investigations have a timeline. The customer has a time frame and a cost allowance in which the investigation is to be completed, and the timeline fits into this time frame. Figure 7 is one such timeline.

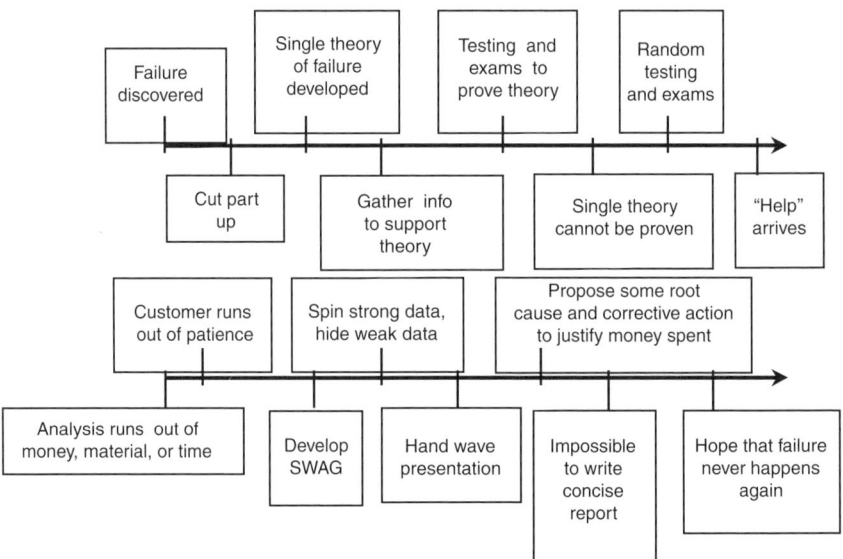

Fig. 7 Timeline of an *unorganized* failure investigation

It is a timeline of a failure investigation in which there is very little organization and planning.

According to Fig. 7, the failure is discovered and the first action taken is to cut up the part. While the part, along with any evidence on it, is being flooded with coolant during cutting, a meeting is held with all possible experts concerned and a single theory as to the failure's root cause is developed. Sometimes there is no meeting, because the engineer thinks that he or she personally can handle everything.

Next, all the possible information to support the one theory is gathered, reviewed, and summarized. Exemplar testing and evaluations such as metallurgical, chemical, and so forth are performed to prove the theory. Suddenly, it becomes clear that the theory cannot be proved. Either too little information and data have been collected to support the theory or too much information and data have been collected that support a different theory.

Now the fun begins. Random testing and evaluations begin on the residual pieces of the failure in an attempt to close in on the root cause. However, many of the samples have been destroyed or contaminated by previous tests and are now useless. The information and data are reevaluated for other possible leads. Meetings are quickly held on how the failure investigation is going to "get back on track."

At some point in time, usually about two to four weeks into the failure investigation, "help" arrives unannounced. This is not the good kind of help, such as more money or more labor. This help is in the form of management who will now require daily updates and charts, adding more workload to the engineer's already busy schedule.

Sooner or later, one of two things happens. The customer runs out of patience, or the failure analysis runs out of time, money, or material to test and examine. An answer must be provided, so a "SWAG" is developed (i.e., a "Scientific Wild-Ass Guess," a highly technical term). The word comes down to spin the strong data and hide the weak data. A presentation is prepared that has all the substance of a magician performing sleight-of-hand tricks. The purpose is no longer to determine the root cause of the failure, but to justify the money spent on the failure investigation. A root cause is proposed, but not well supported by evidence.

Lastly, a concise report is impossible to write. There is no log of the direction of the failure investigation over time, or the various decisions and conclusions made along the way. Everyone just hopes that the failure never happens again.

Sound familiar? I hope that you have not run a failure investigation like the one just described—but you may well have been part of one.

Now look at the failure investigation timeline in Fig. 8. In this investigation, the failure scene is first documented with hundreds of digital photographs and video. All related hardware and parts are placed in a safe, benign environment to be cataloged and preserved. The information-gathering process starts by obtaining drawings, maintenance records, and so on.

The failure investigation team is now formed. The team consists of all technical experts who are involved with the part and, if possible, the customer and the supplier who made the part. The first and best approach for a failure investigation team is to exclude no one of value to the investigation. However, this can be much more complicated than it seems.

Who do you want on your team? The failure itself will many times direct the size of the team. The obvious choices are customers, subject matter experts, and suppliers. But sometimes customers and suppliers create more problems than they are worth. Sometimes the team's formation can be based on two questions:

1. Who do you *need* on your team? Who adds *value*?
2. Who do you *not want* on your team? Who does *not add value*?

As noted earlier, the obvious answer to question 1 is the customer, subject matter experts, and the supplier. The obvious answer to question 2 is management, at least at the beginning. But the political situation may not allow this decision.

Formation of the team is important because the number of people on the team increases its complexity. A certain number of team members allow things to work well; more than that, and things start to decline. You reach a point where there are diminishing returns, and nothing is getting done. If you see this happening it is time to end the meeting, because you are wasting time and sooner or later everyone is going to figure that out.

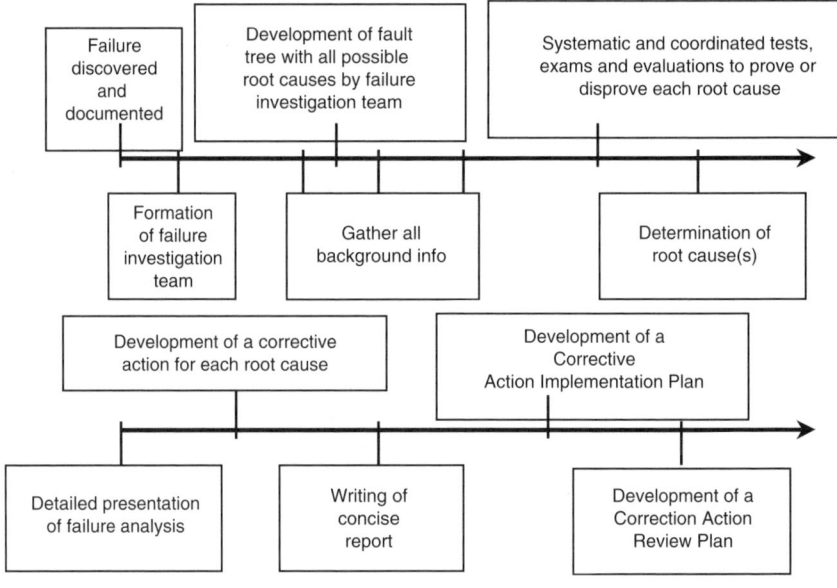

Fig. 8 Timeline of an *organized* failure investigation

If this happens and you, as the team leader, do not address it, the team members may begin to think that the team was a bad idea, that things aren't starting out right, and they will lose confidence. As the team leader, it is your responsibility to consider team dynamics to accomplish the established goals.

A potential solution to maintaining control of a failure investigation team is delegation of work and creation of subteams. One choice is to break into subteams by organizational groups—management and technical personnel, for example. Timing the introduction of certain members onto the team also can help maintain organization and coordination.

The first order of business is to develop a fault tree, with all possible root causes listed and categorized. At the same time, all possible information about the part and how it was designed, fabricated, and serviced should be gathered. This information also includes all operational history.

Using the fault tree as a guide along with good engineering judgment, a systematic and coordinated set of exemplar testing and evaluations (e.g., metallurgical, chemical, etc.) is designed and performed to either prove or disprove each root cause. Disproving a root cause is just as good as proving a root cause. The objective is to prove or disprove every root cause. Some tests or evaluations will be more advantageous or crucial than others if they can prove or disprove multiple root causes. The order and sequence of these tests and evaluations will take some planning.

The result of this well-organized and well-planned failure investigation will be the determination of one or more root causes. The "help" described previously may or may not arrive but will usually go away once they discover that the investigation is well organized and "on track."

The notes from this well-organized failure investigation easily provide enough information for a detailed presentation. In fact, the presentation is an ongoing part of the failure investigation, changing as the investigation proceeds. Finally, a concise and well-written report is compiled, which is a valuable asset to the company.

Now here is where the two timelines differ the most. Instead of hoping the failure does not ever happen again, the failure investigation team immediately develops a Corrective Action Implementation Plan. Various corrective actions for each root cause are suggested and, if necessary, developed or validated with testing. These corrective actions are then implemented.

Lastly, the team develops a Corrective Action Review Plan. After the failure investigation it is common for corrective actions to be implemented, but it is rare that any plans are made to follow up and see if the corrective actions are working. It is possible that the true root cause was not discovered and that the corrective action doesn't work. Therefore, part of the Corrective Action Review Plan involves creating a timeline for checking the effectiveness of the corrective actions.

REFERENCES

1. Air Safety Online, aviation accident database, www.crashdatabase.com (Web site accessed December 2004)
2. Helicopter Crashes, Willis & Associates, www.helicopter-crashes.com (Web site accessed March 2005)

Case History: Firestone 500 Steel-Belted Tire Failure (1972–1978)

James M. Flammang

Seven and a half million Firestone-built tires were recalled when it was discovered that their treads were likely to separate from the main structure, especially when driven underinflated.

Background

After a promising send-off in 1972, Firestone's new "500" steel-belted radial tires ran into trouble. Customers began to return failed tires in unusually high numbers, while tragic accidents were attributed to blown-out Firestone 500s. Most common of the complaints was that the outer tread—the portion of the tire that hits the road—was inclined to slither apart from the rubber carcass or main body.

Through the early decades of motoring, tire technology lagged behind other automotive developments. Not until the mid-1950s did the first tubeless tires arrive. Then came fiberglass-belted tires, adding strength to the carcass. Domestic cars soon borrowed the radial belt pattern from European automakers, and it began to elbow aside the bias-belted styles. Radials have their plies or inner core material running at right angles to the beads or sides of a tire, giving better strength, traction, and fuel economy.

By the 1970s, the new steel-belted radials seemed to promise the best of all possible worlds, mixing the strength of steel with the stability and longevity of the radial belt pattern. Industry sales began to sag by 1974, partly because radials lasted longer, creating less demand for replacement rubber.

In October 1973, *Consumer Reports* rated the Firestone 500 steel-belted radial in the top two. Neither the 500 nor any other tire showed damage in a 325-mile test at 90 miles per hour, so the magazine based its evaluation on tread life, not safety factors. No one seemed to suspect that anything within the tire might cause trouble as it aged.

Actually, someone within the Firestone organization did fear trouble almost from the start. In September 1973 (a year after production began), Thomas A. Robertson, the director of development, sent a memo to top management warning, "We are making an inferior quality radial tire which will subject us to belt-edge separation at high mileage." Nevertheless, Firestone made more than 23 million such radials over the next five years, and insisted the tire had no safety defects.

In 1972, Robertson had expressed concern about Firestone's relationship with automakers that installed equivalent tires (made by Firestone, virtually identical to the 500s, but lacking the Firestone label) as original equipment on new cars. "We are badly in need of an improvement in belt separation performance," he warned, "particularly at General Motors, where we are in danger of being cut off by Chevrolet because of separation failures."

In one of a series of incidents, the Louis Neal family was driving near Las Vegas in June 1974. When one of its Firestone 500 steel-belted radials blew out, their car went out of control and crashed. The parents were killed, and one child severely crippled. Firestone settled out of court for $1.4 million.

Between 1973 and May 1978, when the House Subcommittee on Oversight and Investigations (Fig. 9) opened hearings, 15 deaths had been

Fig. 9 Reps. John Moss and James Collins of the U.S. House Commerce Oversite and Investigations subcommittee examine a damaged Firestone 500 tire during that committee's 1978 hearings on the safety of the tire

reported in which such blowouts were a major cause or main contributing factor. By mid-1979, at least 41 deaths were known to have occurred. The federal government counted 14,000 complaints from consumers.

Consumer advocates knew about the Firestone problem by 1976, having received numerous complaints about tread separations and resultant blowouts. In 1977, the Center for Auto Safety—a watchdog group operated by activists who had worked with consumer advocate Ralph Nader—pushed for government investigation. Having discovered that half the complaints it studied involved Firestone, Clarence M. Ditlow, head of the center, wrote to Firestone president Mario A. Di Federico in November 1977. Rather facetiously, he suggested the "company should shift half of its advertising budget into quality control."

Acting upon data supplied by the Center for Auto Safety, the National Highway Traffic Safety Administration (NHTSA) surveyed 87,000 owners of new cars, asking if they had complaints about radial tires. Of the 2226 owners of Firestones who returned a questionnaire, 46 percent reported problems. The rate for Goodrich was 33 percent; Goodyear and Uniroyal, 32 percent. Despite a court order (obtained by Firestone) to suppress the report, these figures were "inadvertently" released, according to the agency.

Even before the large-scale 1978 recall, Firestone 500 steel-belted radials had been subject to four smaller recalls for manufacturing defects, which affected a total of 410,000 tires. But Firestone still opposed a full recall. Instead, it aimed the blame at a careless public, charging that drivers failed to keep tires properly inflated, bumped them against curbs, overloaded their cars, and drove too fast.

The NHTSA investigated for seven months, studying 6000 consumer reports. Then, in July 1978, it made an "initial determination" of a safety-related defect.

On December 5, 1978, Firestone and NHTSA agreed to a recall of the affected tires, built from September 1, 1975, to the end of 1976 (for five-rib tires) or through April 1976 (for the seven-rib style). "Failure," said the NHTSA announcement, "can result in a loss of air with a possible loss of control of the vehicle which may result in vehicle crash." The official decree covered 14.5 million tires, but only an estimated 7.5 million that remained on the road wound up entitled to free replacement. Approximately 6 million earlier tires could be exchanged at half price.

Production of the 500 continued into April 1978, but the recall did not apply to tires made after the cutoff dates, because internal construction had changed by then. Neither did it apply to nonradial (bias-belted) 500s.

Not all of the affected tires wore a "Firestone 500" label. A dozen others included the Firestone TPC (used on GM cars); Montgomery Ward Grappler; Seiberling RT 78; Atlas Goldenaire II; Caravelle Supreme; Shell Steel Radial; K-Mart Steel Radial 400; and Zenith Supreme.

In a departure from customary practice, the federal government listed no specific engineering cause for the flaw and recall. Instead, NHTSA based its decision upon the simple fact that, based on the number and nature of complaints from consumers, an excessive number of these tires had been demonstrated to be defective. Thus, a pattern of failure was deemed sufficient to prompt the recall.

Details of the Defect

Like other belted tires of the 1970s, the Firestone 500 radials carried a set of reinforcing belts beneath the tread. Whether made of fiberglass or steel, these belts encircled the tire's carcass, enhancing the rigidity of its structure. In the case of the Firestone-built radials, heat buildup within the tire appeared to cause the tread to separate from the steel-belted inner layer. When that occurred, the ultimate result could be a blowout.

Prior recalls of some 500s suggested that a serious problem lay waiting. In March 1976, a small number (200) of Firestone 500s were recalled because of a possibility they were "undervulcanized, which could lead to separation in lower sidewall area." A year later came a record-size recall of 400,000 tires, because they exhibited "distortion or separation in tread area" in a high-speed indoor test. Firestone cited a temporary production problem as the reason for the defect.

One technical detail is certain: steel does not bond readily to rubber. Even when coated with brass, the twisted steel wires that form a tire's belt pattern resist becoming one with the carcass. In the early days, all domestic tire makers were having trouble producing radials.

Because NHTSA failed to specify the precise nature of the defect, and no one else presented conclusive evidence, any analysis must be speculative. One such scenario, based upon investigation by a Georgia newspaper, was presented in *Consumer Reports*.

After examining internal Firestone documents and considering basic manufacturing processes, the *Atlanta Beacon Journal* determined that time was a significant factor in tread separation. Most such tires failed later in life. The tire could be trouble-free for thousands of miles, but a foundation for failure may have existed from the start.

To help overcome the fact that the brass-coated steel belts don't bond easily to the tire's rubber carcass, the belts were first coated with a rubber compound. This created a better base; but if that compound absorbed moisture from the air, it could set the stage for possible disaster. When the tire was vulcanized (heat and pressure applied, to harden the rubber), ammonia formed in the base rubber. Mixed with the internal moisture, this created a corrosive substance. Over time, that substance tended to corrode the brass coating of the steel wires, weakening their bond with the surrounding rubber.

A Firestone document from mid-September 1975 appeared to affirm that diagnosis. Employee P.F. Murray wrote that degraded adhesion resulted from "corrosion of the brass caused by ammonia from compounding ingredients combined with moisture." Firestone was investigating "dezincification" of belts and the effects of ammonia and moisture.

Critics charged that even if underinflation were actually the cause of failures, the company was negligent in failing to warn owners of the danger. Underinflation causes a tire to flex far more than normal. In addition to increased mechanical stress on the rubber, it can cause a tire to run hotter. If motorists had been aware that a tire inflated to appreciably less than the recommended volume would trigger eventual tread separation, they may have kept closer watch on tire pressures.

Firestone and its defenders continued to dismiss the charge of an inherent defect. Malcolm Lovell, chairman of the Tire Industry Safety Council, concurred with Firestone that the problem stemmed from underinflation.

Corporate records turned over to NHTSA, however, suggested that Firestone was hardly unaware that trouble was brewing. In addition to the Robertson correspondence, a 1973 letter from Atlas Tire—one of the companies that marketed a Firestone-built radial under its own name—stated that "Firestone is coming apart at the seams and drastic action is required." In 1976, representatives of Shell indicated they were ready to stop marketing their private-brand tire because of customer returns. Montgomery Ward warned that returns of its Grappler 8000 had reached "epidemic proportions," which "amplifies the fact we were given a bad product."

No other radial-tire line had a failure record anywhere near that of the 500 series. Firestone's claimed 7.4% "adjustment rate" (the number of tires returned for refund, divided by the total number sold) was twice the figure for competitors. Goodyear, Firestone's top rival, rated 2.9 percent.

When Representative Al Gore questioned that rate during congressional hearings, Firestone's counsel explained that 500s were their most expensive tire, so customers would be more likely to complain. Further investigation revealed that in reality, an industry record of 17.5 percent (not 7.4 percent) of these tires were returned to dealers.

Impact

Even if Firestone was not culpable for flawed manufacturing processes, the company's angry denials of responsibility, and of any inherent flaw, drew fire in the press. Worse yet, on learning that a recall was imminent, Firestone sold off the final 500s in southeastern states, at clearance-sale prices. *Business Week* described this last-ditch maneuver as having the appearance of a "desperate effort to unload damaged goods."

At congressional hearings in May 1978, Firestone argued that the tire had been discontinued 18 months earlier, to be replaced by a new 721 series (also steel-belted radials). While the phaseout had indeed begun by 1977, and significant design changes were instituted between 1976 and 1977, final 500s were coming out of the factory as late as April 1978.

In a vain effort to block a total recall, John F. Floberg, Firestone's vice president and general counsel, attempted to suppress release of NHTSA's survey of radial-tire owners by obtaining a court order. (Firestone condemned the survey's validity, because half of the questionnaires went to Firestone tire owners.) That effort went awry, as the survey was released anyway, and the publicity told consumers more about the issue than they would otherwise have known. Firestone chairman Richard A. Riley, meanwhile, continued to insist the 500s were safe, and that the company had "been completely aboveboard."

During ten months of debate, noted *Business Week,* "additional revelations about the tires hardened NHTSA's position, as well as public opinion." Firestone's prior knowledge of the "unusually high volume of its customers' complaints" became evident, while the company "repeatedly tried to thwart investigation." *Fortune* magazine charged that in its defense, Firestone had "used tactics that worsened the ordeal," concluding that the company's response "may well become a classic."

Analysts predicted that the wave of negative publicity could skim 2 to 3 percent from Firestone's usual 25 percent share of the replacement-tire market. Meanwhile, the company prepared to spend $230 million for the recall. By mid-1979, about three million Firestone 500s had been replaced, but NHTSA chief Joan Claybrook and other critics expressed disappointment with the pace of the recall.

Firestone still faced a series of lawsuits charging property damage and loss of life, plus a $1.7 billion class action. By the time of the recall, about 250 such suits had been filed. Nine verdicts had gone against Firestone, 22 ruled in the company's favor, and 64 had been settled out of court. Firestone also faced an inquiry by the Securities and Exchange Commission, which wondered if the company had adequately informed its shareholders about the 500 issue. Although the SEC apparently took no action against Firestone, in 1979 a group of about 4000 shareholders filed a class-action suit against the company accusing it of concealing information about the 500 defect. Firestone settled with this group for $3.2 million in 1983, without admitting any fault or liability.

Although some critics went so far as to compare the Firestone 500 debacle to President Nixon's Watergate crisis, the company's woes turned out to be relatively short-lived. To help bolster its faded image, Firestone enlisted folksy actor Jimmy Stewart to help promote the new 721 series. They also offered a then-radical two-year warranty. Before long, Firestone managed to rebound to near its former stature in the industry.

SELECTED REFERENCES

- "The Big Firestone 500 Recall," *Consumer Reports,* April 1979, p 199–200

- "Firestone 500: Blow Out," *The Economist,* October 14, 1978, p 121–122
- S.A. Feldstein, "How Not to React to a Safety Controversy," *Business Week,* November 6, 1978, p 65
- "The Case for Firestone," *Forbes,* November 13, 1978, p 106
- R. Gray, "Firestone Adds Batch of New Entries in Recall Wake," *Advertising Age,* February 5, 1979, p 4
- P. Hinsberg, "Analysts Assess Firestone Recall," *Automotive News,* October 30, 1978, p 2
- H. Kahn, "Secret Documents Hint at Firestone Coverup," *Automotive News,* January 1, 1979, p 1, 23
- A.M. Louis, "Lessons from the Firestone Fracas," *Fortune,* August 28, 1978, p 45–48
- "Motor Vehicle Tire Recall Campaigns," National Highway Traffic Safety Administration, 1976 ed., p 72; 1977 ed., p 79–80; 1978 ed., p 67
- "Forewarnings of Fatal Flaws," *Time,* June 25, 1979, p 58, 61

Case History Discussion

What caused the Firestone steel-belted tire failures in the 1970s? Tires came off, drivers lost control, cars rolled, people died—a scenario that sounds vaguely familiar to what has been occurring over the last few years with tires on Ford SUVs.

The physical cause of the failures was corrosion of the brass coating on the steel belts. When the tire was vulcanized, ammonia formed in the base rubber. When this ammonia mixed with the moisture in the rubber compound, a corrosive substance was created that corroded the brass coating on the steel belts. This led to a separation of the steel belts from the tire carcass after a period of time.

The root cause of the failure was a manufacturing process error in which moisture was introduced into the rubber compound that coated the brass-coated belts.

The Firestone defense was interesting. The company tried to hide facts and then tried to suppress facts. In the end the company had to pay, but it rebounded. The failure appeared to have no ramifications whatsoever. Firestone underwent a short-term loss of business, but within a few years had returned to owning 25 percent of the industry. People have short memories and they bought Firestone tires again. Firestone has strong market recognition and penetration, including the use of Firestone tires by NASCAR.

What will happen today?

If you were Firestone, and you were going through this new failure, would you look back at your business strategy from 1972? Would you follow the same business strategy? It appears the company may be doing just that. They are pointing at the Ford Motor Company and the American public saying the problem is the SUV and careless drivers with underinflated tires, not the tires. They appear to be denying the problem again.

So what should Firestone do? Should Firestone design tires to meet market needs and the public's poor driving habits? Should Firestone design a tire that can be used underinflated, overinflated, driven over curbs, and driven too fast? Should Firestone design to these standards because this is their market? Do you design ships for 100-year storms or commercial aircraft to fly in tornados?

If you answered yes, ask yourself two questions:

1. Would you pay for this tire? Making a tire that would satisfy all these conditions would be very expensive.
2. How much should we as members of the public take responsibility for our own safety, in terms of maintaining proper tire inflation, driving the speed limit, and so on?

Interestingly, there are other corrosion-related problems with coatings breaking down during processing or use and creating problems with materials in a design. One example involves Teflon-coated wires in storage. The process used to coat the metallic wires breaks down the Teflon and produces hydrofluoric acid when stored in areas where condensation occurs. The metallic wires corrode during storage due to the acid formation.

A second example is Teflon seals that break down due to heat or exposure to certain fluids. In some instances Teflon seals, both stationary and rotating, have broken down and caused corrosion of titanium components. Titanium is a very corrosion-resistant material, but is highly susceptible to corrosion by hydrofluoric acid. These reported occurrences indicate that the Teflon breaks down either chemically in the case of the stationary seals or due to heat in the case of the rotating seals. When the Teflon breaks down in the presence of water, it creates hydrofluoric acid. The rotating seal failure was especially surprising because the seals were used in a water system. The designers expected the system to last practically forever, but it stopped operating due to seal leaks caused by acid corrosion/erosion of the titanium. The system lasted only 25 percent of its expected life.

CHAPTER **4**

Nine Steps of a Failure Investigation

NINE STEPS are necessary to the organization of a good failure investigation:

1. Understand and negotiate the investigation goals.
2. Obtain a clear understanding of the failure.
3. Objectively and clearly identify all possible root causes.
4. Objectively evaluate the likelihood of each root cause.
5. Converge on the most likely root cause(s).
6. Objectively and clearly identify all possible corrective actions.
7. Objectively evaluate each corrective action.
8. Select the optimal corrective action(s).
9. Evaluate the effectiveness of the selected corrective action(s).

The first five steps relate to steps 1 and 2 of the four-step problem-solving process discussed in Chapter 3. The next four steps relate to steps 3 and 4 of the four-step problem-solving process. Many failure investigations stop after step 5, but you should try to convince your customer of the value of completing all nine steps. In every report, add a section concerning recommendations (i.e., "This is what you should do") whether or not the customer requests it. If the failure investigation has successfully determined the root cause(s), then these recommendations will already have been formulated in your mind.

Step 1: Understand and Negotiate the Investigation Goals

At its onset, every failure investigation should establish four criteria: (1) the priority of the investigation, (2) the resources available, (3) any constraints imposed, and (4) the goal or goals of the investigation. A discussion should be held to define these four criteria, which combine to immediately set the importance, direction, and expected results of the investigation.

Priority. Always discuss priority first, because this sets the pace and can sometimes determine the resources and constraints. Many engineers do not ask the basic question regarding an investigation's priority, but it determines the completion date and how much time and effort can be expended to achieve that date.

Resources and constraints may simply involve the same three items: money, labor, and time. Clarifying these items ahead of time saves frustration in the long run, especially when trying to get personnel to dedicate time to your project. Constraints can also include the inability to destroy the hardware, lack of similar hardware to review, lack of information, and so on.

Let's discuss the puzzles presented in Chapter 3. Did 60 seconds seem a reasonable time frame to solve each one? Each puzzle arrived in a brown envelope so that no one was able to guess its difficulty. Even though the puzzles are designed for ages two to five, no one usually completes the puzzles in 60 seconds. Surprisingly, no one has ever complained that 60 seconds was insufficient and requested more time. Even more surprisingly, no one complained afterward that the assignment was impossible in the time allotted. The next time you get a failure investigation wrapped in a brown envelope, ask "How much time do I have?" Then negotiate if you believe you need more time.

Goals. Last, but certainly not least, is the failure investigation's goal or goals. Your customer may desire a very simple goal, one that will not require determination of the problem's root cause. Now is the time to discuss goals and the benefits of determining root cause.

In the example with two puzzles in one envelope (Fig. 6 in Chapter 3), no one has ever questioned which puzzle to solve. The person is given a task to perform and attempts to complete it. The puzzles are provided, the time limit is provided, and so is the goal of solving the puzzle. Everyone always attempts to solve both, thus wasting time and money. Even when other people are asked to join in and help solve the puzzles, no one ever asks which puzzle is the goal.

Failure investigators need to gently inform their customers/managers when they are not providing proper direction. This concept is called "managing your manager" or "managing your customer." Customers sometimes provide bad information because they may not truly know what they want. The failure investigator must talk with them regarding expectations and goals, because it is not prudent to spend a lot of time and money on something the customer does not want. In the end, customers and managers will appreciate these questions. You have to negotiate.

For example, remember the trunnion pin fabricated from Custom 455 in Chapter 2. I was requested to determine a minimum property allowable for the trunnion pins that had been received with mechanical properties below the specification limit. The goal of the program manager of this hardware was to provide information for a stress analysis to determine if the hardware could be deemed acceptable on a nonconformance report. I looked at this problem as a failure of the heat treatment or the raw material. Since this situation involved an approved heat treat supplier to our company and we had fabricated many other pieces of hardware from a large amount of raw material in stock, I persuaded the program manager that it would be in the company's best interest to determine the root cause of the problem and investigate the heat treatment and raw material as well. Consequently, the failure investigation would determine whether the company had other hardware at risk or if it was necessary to change a heat treat or material specification in order to better control the heat treatment or raw material. The program manager saw the problem as unique, but I saw the problem as systemic.

One very tough constraint was that I had no hardware for analysis; the program manager wanted to save each and every trunnion pin. During conversations with the customer, however, it became evident that they were interested in some analysis. Therefore, I reviewed the data and discovered that one of the trunnion pins had mechanical properties that were inconsistent with the others. I convinced the program manager that we had one part that was quite different from the rest and that the customer would be happier if we analyzed one of the pins. He agreed, the customer was happy, and I had a part to analyze.

Summary. Remember to negotiate priority, resources, constraints, and goals—primary as well as secondary goals, of which your customer may not have knowledge. Always negotiate these four criteria at the beginning of the failure investigation so that all parties are in agreement. And once the negotiation is completed, document these agreed-on criteria in a one-page memo. There is nothing worse than when a customer asks an engineer to find out what time it is and then is given a report on how to build a clock. On the other hand, it is also unacceptable when a customer asks why the clock stopped (root cause), only to be told the simple answer of what time the clock stopped.

The emphasis I place on this initial negotiation and documentation between the customer or manager and the engineer in charge of the failure investigation is best explained by the parable that follows.

Know Your Customer: The Great Captain, the Good Captain, the Bad Captain, and the Terrible Captain. On the night of April 14, 1912, the White Star Line's newest and largest luxury steamship, the *Titanic*, struck an iceberg during its maiden voyage from Southampton, England, to New York and sank in less than three hours. More than 1400 people lost their lives. Of the newest watertight design, the *Titanic* was considered "unsinkable." Imagine that you are the chief engineer and designer of the *Titanic*, aboard to provide any technical expertise the captain may require.

As the technical expert you know that the *Titanic* is operating in shipping lanes known to have icebergs and that April is one of the worst months for them. You check the ship's logs and discover that four icebergs already have been spotted by lookouts during the first five days of the voyage. On the night of the fifth day, April 14, the sea is exceptionally calm and the night is moonless, factors that make visually spotting icebergs difficult. You are outside enjoying an after-dinner walk and notice that the ship is moving very fast. A quick walk to the bridge and you discover that the ship is moving at the fast speed of 21 knots. The bridge crew informs you that the captain is trying to set a speed record on the ship's maiden voyage. You approach the captain and recommend that the ship be slowed down until morning because of two known facts: Icebergs have been spotted earlier, and the current calm, moonless conditions hinder spotting them at night.

Here is where we learn about the great captain, the good captain, the bad captain, and the terrible captain. The great captain would heed the technical expert's warning and slow the ship until morning in order to avert a potentially disastrous situation. The good, bad, and terrible captains would thank you, but ignore your recommendation and do nothing. The good, bad, and terrible captains would each inform you that he has been a captain for many years with a lot of experience in these shipping lanes, that the ship's sixteen watertight compartments make the *Titanic* unsinkable, and that he has promised the White Star Line management a speed record.

As history tells us, the 46,000 ton *Titanic* strikes an iceberg at a speed of 21 knots at 11:40 P.M. and the energy absorbed is so insignificant that the impact is hardly felt by the passengers and crew. As the technical expert, you gather data to assess the vessel's integrity and determine that six of the sixteen watertight bulkheads have been breached and are filling with water. It is your opinion that the ship will sink. You don a life vest, grab another one, and find the captain.

You tell the captain that the ship is going to sink, offer him a life vest, and recommend that he start to evacuate the ship. At this point in time, the good captain would learn from his previous mistake, heed the recommendation of the technical expert, don the life vest, and order evacuation. The bad and terrible captains would say that you must be mistaken because the *Titanic* is unsinkable. They would refuse the life vest and not begin an evacuation until they were certain the ship was sinking. These two captains would also command you to keep this information to yourself and not alarm the passengers.

The order to evacuate is given at 12:20 A.M. As the technical expert, you know the *Titanic* has sixteen lifeboats and four emergency crafts that are capable of holding approximately 1200 of the 1900 passengers on board if the recommended number were put in each lifeboat and craft. This design deficiency was due to the number of required lifeboats being based on ship tonnage instead of the number of passengers, and on the watertight design of the *Titanic*. During the evacuation, you notice that the officers are only partially loading the lifeboats due to the fear that the lifeboats will buckle or that the davits, or cranes, lowering them into the water will break. You know that the lifeboats and the davits have undergone testing and that they can easily carry the required number of passengers. You also know that the testing was performed using male passengers. Therefore, the lifeboats can be overfilled considering the number of women and children currently on board. You also notice that the distress rockets have not been fired to alert other ships to the emergency aboard the *Titanic*.

You approach the captain a third time and recommend that the lifeboats be filled slightly beyond capacity and that the distress rockets be fired. The bad captain would finally heed the technical expert's recommendation and instruct the officers to overfill the lifeboats and fire the distress rockets in an attempt to save as many people as possible in this desperate situation. The terrible captain would say that you may know about the design and operation of a ship, and that you may be able to assess the damage to a ship, but that you know nothing about evacuations. The officers shall trust their own judgment, instruct the lifeboats to come alongside the *Titanic* cargo ports to pick up more passengers, and the distress rockets will be fired at the appropriate time. The terrible captain would then tell you that your advice was no longer needed and to get in a lifeboat and get off his ship.

The *Titanic* breaks in half and sinks at 2:20 A.M., only two hours after the evacuation began. The one ship to respond to the *Titanic* SOS was the *Carpathia,* which was 58 miles away and arrived at 4:10 A.M. The *Carpathia* rescued only 711 people from the *Titanic* lifeboats. The cargo ports were never opened, so many of the lifeboats floated away only partly filled. The *Californian* was only *20 miles* away, but its radio operator had gone off duty shortly before the *Titanic* sent out its S.O.S. At this distance, the *Californian* would have seen any distress rockets and could have arrived much quicker. Approximately 500 people died needlessly because of the lifeboat loading and distress rocket mishaps.

You are one of the lifeboat survivors looking for people in the water during the two hours between the sinking of the *Titanic* and the arrival of the *Carpathia*. As the technical expert, you know that a person is not going to last long in freezing cold seawater at -2 °C (28 °F). The lifeboat comes across the terrible captain. He is holding onto some wreckage, while clutching the ship's heavy bell and logbook.

For the last time, you as the technical expert advise the terrible captain to drop the bell and logbook and to get out of the freezing water and into the lifeboat. The terrible captain tells you that it is his duty to save these items. He looks at you and says, "You were my technical advisor. Why did you not help me to avoid this disaster? And once the disaster occurred, why did you not advise me of the best course of action?" The terrible captain then sinks below the surface of the water, still clutching the ship's bell and the logbook.

You sit back in the lifeboat, stunned at the terrible captain's response. You then look up at your lifeboat companions. They are staring at you in anger and disbelief. Finally, one young boy asks, "Sir, was it in your power to avoid the ship sinking? Could you have saved my mommy and daddy? Why did you not help the captain when he needed you?" You look at the young boy and the others and realize that there is no explanation they will believe at this point in time.

The point of this story is to underline the importance of the initial negotiation and documentation of the communication between the customer or manager and the engineer in charge of the failure investigation. The importance of this initial step varies depending on whether your customer is a great, good, bad, or terrible captain.

Step 2: Obtain a Clear Understanding of the Failure

What is the problem? This question is the first step in the four-step problem-solving process. This should also be the first question asked in a failure investigation. What happened? Why are we here? Engineers so often want the details, they forget to hear the reason their customer wants their help. At this point, information is crucial.

Next, you need to gain as much expertise as possible about the component, machine, or system. If you do not understand what you are trying to investigate, you will never get to the root cause. Does the failure involve one distinct part or is it part of a widespread, systemic problem? This often is a question the customer wants answered, but is afraid to ask. You should ask this question.

Many tools are at your disposal, and brainstorming is one of the best. Get together everyone—and I mean everyone—and brainstorm theories as to why the failure occurred. Information pours out in a brainstorming session, and you may learn some very important facts. The drawing may call out one process specification, but the people on the shop floor may have replaced that specification a few months ago through production documentation. This is important information that would be missed by merely reviewing the drawing. Review all documentation and records available on operational history and fabrication. Time consuming? Yes. Tedious? Yes. Important? Definitely yes! Remember, the devil is in the details. Be rigorous in all aspects of your investigation.

During every failure investigation, I remember this quote:

> It is not only in the finding that we learn much, but also in the looking. All things, big and little, teach us; perhaps at the end, the little things teach us the most.

Those words were spoken by Dr. Van Helsing during his search for the elusive Count Dracula.

What do I mean when I say talk to everyone? For example, how would you discover the use of Teflon grease in the test areas of a shop? During testing of assembled hardware, parts do not always fit together or operate as well as they should, so technicians in the test area sometimes use Teflon grease and everything works fine—although the grease is not allowed per any drawing or test procedure. There is only one way to discover its use, and that is to spend time in the test area and talk with the technicians as to how they performed the testing. Just reviewing the paperwork will not provide the answer.

For many applications a little Teflon grease is not a problem. Many technicians in test or shop areas come from other industries where using such grease is perfectly acceptable. However, the aerospace industry, especially when involved in space applications, is very restrictive on aerosols and greases because of outgassing requirements. Most of the greases commonly used on a shop floor will outgas when put into space. The outgassing material adheres on the surfaces of sensitive hardware such as solar panels or heat rejection panels, accelerating degradation. The International Space Station (ISS) is designed for a lifespan of 20 years. If the solar panels that support the life-support systems degrade in five years, then the ISS lifespan is shortened.

The failure scene also is very important. Remember: "Document and protect." If you cannot visit the actual failure scene, request drawings and/ or photographs of it. Photographs of the failure scene both before and after the failure are frequently helpful. If you are able to examine the scene, bring a field investigation kit and document everything. Photographers have an old saying, "film is cheap." With the advent of digital cameras, this can be rephrased to "jpegs are free." Take hundreds, or even thousands, of photographs. In addition, request and look for video cameras, used for either process control or surveillance. All information is valuable.

Why take so many photographs? A common children's memory game provides the answer. Often found in the Sunday newspaper comics or in children's books, the game's goal is to remember the objects depicted in a photograph.

Study the photograph in Fig. 1 for 60 seconds. In 1946, during the U.S. Operation Crossroads, the aircraft carrier CV-3 USS *Saratoga* was sunk by the blasts produced by the Able and Baker nuclear explosions. The photograph of the carrier's bridge in Fig. 1 was taken in August 2003 by the author at a depth of 110 ft off Bikini Atoll in the Marshall Islands of Micronesia.

Now, without looking at the photograph, answer the following questions:

- How many portholes are shown?
- How many portholes still have blast covers in place?

- How many tables are there?
- What time does the clock show?
- Is there a lamp on the table?
- If so, what kind of lamp is it?
- How many white cups are on the table?
- Is there a communication tube on the bridge?

The objects in Fig. 1 range from obvious to subtle. So will the details of a failure investigation. Hopefully, enough information will be captured in failure investigation photographs to document all the information that is there to find.

The point of this exercise is that having the photograph to review later makes it easier to discover or remember any hidden objects. It would be impossible to discover or remember these hidden objects based on written or verbal accounts. A picture is definitely worth a thousand words, if not 10,000. And a photograph is much better than your memory.

Field Investigation Kit

When examining the scene of a failure, a useful tool is a field investigation kit (Fig. 2 and Table 1). Let's expand on some of the items listed in Table 1.

Fig. 1 USS *Saratoga* bridge. Photograph by Daniel P. Dennies, Bikini Atoll, 2003

An open and questioning mind is your most important tool. As an engineer performing a failure investigation you must be professional, questioning, pleasant, and businesslike.

A Good Attitude. It will do you good to remember that you need the personnel at the failure scene more than they need you. When you walk into the site where the failure occurred or the hardware was fabricated or heat treated, you need to find out what the people there saw, what they heard, and what was going on. Be pleasant. They have firsthand information about what they saw or heard. Many times they will not know that what they saw or heard is important. But you will.

For example, I once walked into a machine shop to discuss some problems we were having with a material. I struck up a conversation with the machinist operating the single-point lathe. I told him I was there to investigate a troublesome part and he said, "I remember that part. When I was machining it, I could hear a *click click click click*—very strange." What he heard was the effect of a martensitic precipitation-hardening (PH) stainless steel that had too much delta ferrite in it. A small amount of delta ferrite, from 2 to 5% maximum, is allowed in martensitic PH stainless steel. The delta ferrite is a secondary phase that is a result of the chemical composition. It is quite soft. What the machinist heard during the machining operation was the tool moving through pockets of the softer delta ferrite and then making a clicking sound as it hit the harder martensitic structure that should represent 95% of the material. To him it was just a click, but to me it provided information that the original bar stock must have contained more than the 5% maximum amount of delta ferrite. By the time I received the part, most of the material had been machined away, so the delta ferrite content did not look too bad. There was no raw stock left to examine. All my information came from the machinist. He did not know the technical reason for the clicking, but he provided a lot of good information.

A professional demeanor is very important. If you look professional and organized, people will respond. Make sure you have the correct tools when you arrive. If you come prepared with the proper equipment, the customer will realize that you know what you are doing. If you look like

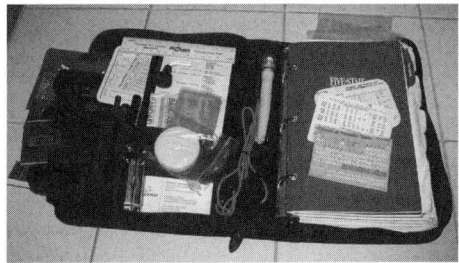

Fig. 2 Field investigation kit. Not shown are digital camera, cellular phone, and laptop/notebook computer.

you know what you are doing, the customer will be more attentive and helpful.

My repeat customers know how I work, what process I will follow, and that we will discuss the expected result. They come prepared with information as to how much time and money are available and the project's priority. They also have a reasonable expectation of the result. Consider your consistently professional and organized demeanor as a means of cross-training your customer. The first couple of times it can be very painful, but people start to know how you operate. The results will speak for themselves. You may have to go through this cross-training a few times, but the customer will see the benefit of the process and accept the way you operate.

Table 1 Field investigation kit contents

Item		Reason
1	Open and questioning mind	Be prepared. Ask questions. Be ready for the unexpected.
2	Good attitude	You need the people you meet more than they need you.
3	Professional demeanor	If you look like you are organized and knowledgeable, people will consider your presence important and may offer more help.
4	Laptop/notebook computer	Spreadsheets, documents, etc. PDA may suffice.
5	Digital camera	Take as many pictures as you can of the item, the area, supporting equipment, people, anything and everything. Be careful about color and document your location when taking a photo.
6	Known color chart, white piece of paper, etc.	Kodak gray or color chart, hardware paint charts, red button, your tie; anything that later can be used as a standard to judge picture color
7	Scales or rulers: steel and plastic	Use in photos for scale: gray steel rulers on light backgrounds, white plastic rulers on dark backgrounds. Steel rulers will indicate whether something is magnetic; plastic rulers may work better if the subject is magnetic.
8	Tape measure	Measure long distances.
9	Magnets: flat one and a "wand" with an extension	Identify magnetic and nonmagnetic materials. Use the wand to retrieve items or to collect magnetic debris; immediately separate debris into magnetic and nonmagnetic.
10	Loupe and magnifying glass	Inspect samples. Loupe is higher magnification ($10–25 \times$); lower-power magnifying glass can also be helpful.
11	Indelible ink marker (fine)	Mark items, bags, bottles.
12	Flashlight	Look at items in dark areas, etc.; 90° bend attachment allows inspection of holes, and pinlight attachment allows inspection of small crevices.
13	Conductivity tester	Check surface conductivity or nonconductivity.
14	Mirror	Check around corners and under objects.
15	Surface finish comparators	Verify machined, cast, electrical discharge machined surface finish standards.
16	Leatherman tool or pocket knife	Scissors, screwdriver (flat and Phillips), punch, tweezers, blades, plastic toothpick; can't be carried on an airplane. Uses include examination tools, surface finish profilameter, cutting, etc.
17	Pen and pencil	Just because you are prepared.
18	Sheets of blank and lined white paper	Color standard, note paper, drawing paper, collection funnel
19	Plastic bags, 100×100 mm (4×4 in.)	Sample collection
20	Swabs	Sample collection; always keep one for control sample
21	Alloy reference list	Materials, compositions, specs, data
22	Hardness conversion charts	Martensitic and austenitic charts
23	Addresses and phone numbers	Assistance, info, etc.
24	Technical information	Specification lists, design criteria, drawings, notes, etc.
25	Cellular phone	Immediate access
26	Eyewitness report forms	Better if these are e-mailed or sent ahead to be filled out as soon as possible.
27	Other	Anything else you think you will need

A laptop or notebook computer is a very powerful and necessary tool. These computers are mobile workstations that can maintain all your records such as specifications, drawings, procedures, past failures, and so forth. In addition, you can communicate with other people or search the Internet. You can write your reports or start your fault tree on a saved template. A personal digital assistant (PDA) may also do the job, provided it has the right software.

Cameras. Documentation via a digital camera or video camera is very important. Take a thousand pictures if necessary, because jpegs are free and no one has a memory that is as good as 50 photographs. Also, once you begin reviewing the photographs, you may see new things that you missed.

A known color chart is a necessity. One problem with digital cameras is that they are so smart that they may alter the photograph, resulting in colors that may not be true all of the time. Therefore, you must have something with a known color in each photograph. The item can be anything—a standard gray scale on a piece of paper, a red tie, a striped scale or ruler—just as long as you know what the color is and can take it home with you so that when you look at the photographs you can decide if the colors are true. Do not use something black, because if the shades in the photograph change, it may be hard to discern the change in the black standard. Use solid and primary colors like red or blue. Find something that is a known color that you will have later in your possession. It is a very simple trick.

Another positive aspect of a digital photograph is that you can quickly alter it to match the standard. However, please note that for legal cases altering digital images used for evidence usually is not allowed. The need for a true photograph is especially important if the failure involves corrosion. Consider that copper or bronze corrosion can be a mix of green and blue, and iron corrosion can be a mix of red and brown. If the photographs do not represent the true colors, it will be impossible to understand what material is involved.

Scales, Rulers, and Measuring Devices. A basic rule taught to all failure investigators is to place a scale or ruler in every photograph. If the photograph is of a large piece of hardware, then ask a person to step into the photograph and note the person's height. Carry a gray steel scale and a white plastic scale in your kit so that you can put a dark scale on a light-colored part and a light scale on a dark-colored part. The fact that one is metal and one is plastic also provides handy variation. A small tape measure is another good tool that easily fits into a field investigation kit, and you may also require a larger tape measure. Tailor the tools to the particular failure investigation.

Magnets can be helpful for material identification, as well as debris collection and separation. If not all the debris is collected with the magnet, then you know there may be two types of materials in the debris. Each plastic bag with debris should be identified as magnetic or nonmagnetic.

Loupes and magnifying glasses are essential tools for a field investigation kit. The loupe is higher magnification ($10\times$–$25\times$). A low power magnifying glass can also be helpful to inspect samples.

A marker with indelible ink for identification is also necessary.

A small flashlight may provide additional light for examination. Attachments allow you to look around corners and into very small cavities.

A conductivity tester can be purchased at any hardware store and is perfect for determining whether the protective coating on hardware is still intact or present at all.

Mirrors are very handy, especially when trying to look around corners in hardware.

Surface finish comparators are metallic standards produced by companies like GAR that simulate various electrical discharge machining (EDM), casting, and machining surface finishes. They provide a standard for various root mean square (rms) surface finishes for different kinds of fabrication operations. For example, if the casting drawing states 125 rms, but you look at the standard for castings and determine the surface finish is 500 rms, that is a valuable piece of information. These standards are a good reference point and make the observation of surface finish more fact than opinion. You can become pretty good at determining surface finishes by eye and feel, but nobody argues if you have a standard. And you can put it in the photograph with the scale.

Leatherman Tool or Pocket Knife. My grandfather used to say that no gentleman leaves the house without a pocket knife in his possession. For engineers performing failure investigations, pocket knives (or the popular Leatherman devices) are valuable for their cutting, scraping, and sampling capabilities.

Other Tools. Your field investigation kit will grow in size with time and change with the type of failure. Many times you will not know exactly what tools are required. But those tools mentioned so far should always be in the kit. A few more basics include a pen and pencil, sheets of blank and lined white paper, 100×100 mm (4 by 4 in.) plastic bags, and swabs for debris collection.

The field investigation kit should contain basic information references such as an alloy reference list and hardness conversion charts. Other important tools include an address book with phone numbers and a calculator, both of which are part of any PDA. The kit should contain any pertinent technical information regarding the failure, such as specifications, design information, and company procedures.

A cellular phone seems to be an everyday item for every person on the planet these days and is a handy tool for calling coworkers, industry experts, laboratories, raw material suppliers, and so on. The kit should also contain eyewitness statement forms.

A field investigation kit must be tailored to the work the failure investigator is performing. Other items might include dental tools or plastic knives for scraping. A small borescope would make a great addition, if required. Make it your own. What do you need for the failures common to your business and industry?

Eyewitness Statements

Equally important are eyewitness statements or incident reports. Unfortunately, when things go wrong, eyewitnesses seem to disappear or, at best, are reluctant to speak. Sometimes they are shielded by management or unions. Getting information from them is a true talent. I suggest you send standard forms ahead of time to be filled out by eyewitnesses. The number of statements that are returned and the completeness of each will provide some indication of the level of cooperation you can expect. In many cases, you may never actually see the failure scene and will have to depend on available photographs and eyewitness statements. Good information can be obtained from the people who witnessed the failure, especially if it occurred at a different supplier or a different plant. The name of the form is not important—an eyewitness statement, an event report, an incident report, or an investigation report. During a failure investigation they are good things to have, but very difficult pieces of information to obtain.

Talking with Eyewitnesses. In every failure investigation you may not be able to get to the person who knows what happened. Here is a simple example. ISS truss segments are large hexagons that measure 4.6 m (15 ft) across and more than 12 m (40 ft) in length. Each truss was held off the ground during assembly by tooling at four points, two at each end. One day an accident occurred while moving one of the trusses. The moving procedure was to attach the overhead crane to the two points at one end of the truss and adjust the crane until all slack in the crane was removed. Then a second overhead crane was attached to the two points at the other end and adjusted until all slack in the crane was removed. The truss would then be disassembled from the tooling and moved. Unfortunately, the two overhead cranes were operated from one control. On the day of the accident, a technician set the first overhead crane to the proper elevation, then turned his back to that crane and started to adjust the second one. But he used the wrong part of the control and was actually continuing to raise the first overhead crane. There was a loud bang as the tooling fixture broke and the truss sprang up from the tooling.

The entire ISS engineering team was contacted and told to meet in the fabrication area. The team immediately requested to talk to the technician operating the overhead cranes, but the union stewards had already taken him out of the building. He was unavailable, gone, and the union stewards related to us what he told them had happened. The team pieced the story together based on information from the union stewards and other personnel in the fabrication area, but it took five days because the union was shielding the technician.

Remember, you may not always be able speak with the eyewitnesses to a failure. Such incidents may involve politics, fear, anger, unions, job security, and so forth. Can you get to the truth? Yes, but it may take some work.

Investigation Team Composition. When speaking to eyewitnesses and visiting the failure site, how many people should you send and exactly whom should you send? Sometimes it is best to arrive alone. This is less obtrusive and intimidating than arriving with a large team. Eyewitnesses may be more willing to speak to one person who is just trying to find out what happened. While this is a good approach, the problem is that you only have one set of eyes and ears and thus may miss something. A two-person team has double the eyes and ears.

What is the size limit for a team? At what point does the team become too many? At what point does the team become perceived as a lynch mob and scare the eyewitnesses? The moment the eyewitnesses are frightened is the moment you will no longer be provided information. That number is different for each failure investigation and depends on several factors, including the investigation's size and complexity, the size of the plant to be visited, and the number of eyewitnesses to be interviewed. If a large team is required, then sometimes it is smart to split up and address separate areas, such as management issues, design issues, and manufacturing issues.

Cultural sensitivities at an investigation location are another consideration. For example, two engineers from Ohio were sent to a supplier in South Carolina to investigate a failure of some hardware. The two engineers first met with the manager of the facility. During the introductions, the manager learned that the one of the engineers was named Mr. Sherman. The manager looked directly at the engineer and told him that he would not do well at this facility. The engineer was perplexed and asked for a reason. The manager asked Mr. Sherman if he had ever heard of General Sherman's march through the South during the Civil War. Both engineers said yes but could not believe that association would impede the failure investigation. After one week, Mr. Sherman was replaced by another engineer.

Standard Forms. The best approach is to create eyewitness statements in a standard format. Personnel are familiar with forms and usually do not find such documents to be confrontational and intimidating. Even better is to make the eyewitness statement a company procedure, complete with the company letterhead, a procedure number, and instructions on how to fill it out. People will perceive it as just another procedure, rather than a special investigative form or an intrusive inquiry. The eyewitness statement forms should be electronic and searchable. That way they can be quickly e-mailed back and forth. This also will allow eyewitnesses to write everything they can remember because, unlike a paper form, space is not limited. The information also will be useful later for generating statistics or searching for similar failures, failures at the same plant, failures involving the same personnel, and so on.

As noted previously, it is best to send the eyewitness forms ahead of time. The number and the completeness of the responses will indicate the cooperation you can expect once you arrive. If you send out forms to the five people involved in a failure—the machine operator, the manager on duty, the foreman, and two other employees—and you get back one form from the plant manager, this suggests you may not receive much cooperation. Either the manager did not pass out the forms, or the employees refused to complete them. If all the forms are returned, then you can take a quick look and compare them. Do the answers match? Do the dates and times match? The consensus or differences noted in the forms provide good information.

Another reason to send eyewitness forms ahead of time is if you cannot arrive at the failure scene quickly. People's memories fade fast, and studies have shown that eyewitness stories have a tendency to change over time. Memories become "improved" as people tend to change the story to favor themselves or others. Such changes are usually not intentional, but there have been cases where eyewitness accounts have changed to make an employee appear more innocent. In addition, as different eye witnesses discuss the failure, a "collective" memory is established. And, eye witnesses "recall" facts that were impossible for them to have seen.

Questions to Ask. An eyewitness statement form should cover who, what, when, where, how, and why. Get all the basic information you can: name, position, date, time, shift, location, the weather, the type of equipment being operated, the procedure or specification to which the eyewitness was operating, the material, the maintenance record, whether the eyewitness knew anything about the failure history. Ask everything. Some answers will remain blank, but you will be surprised by the completeness of other answers. People know more than you can imagine.

Eyewitnesses may also possess much knowledge about how the plant operates in general and what happened during the failure incident in specific. An eyewitness may be a great source of answers to questions concerning the plant and the shop floor, such as "What is the normal operating procedure?" "Was the normal operating procedure being followed?" "Are you aware of recent changes?" This last question elicits information regarding planned and unplanned deviations. Another specific question would be, "Has this kind of failure happened before?" Put that right on the form. The manager will say no, but the operator may say that it happens about every six weeks. The manager does not always know what is happening on the shop floor. Even worse, many times a company will simply dismiss a failure the first few times it happens and write it off as a unique or one-time occurrence. These occurrences are then forgotten until the next one. Unfortunately, since the first failure is dismissed and undocumented, no one may remember it and so the second failure may also be deemed "unique." It is truly a downward spiral.

Ask not only what happened during the failure incident, but also what happened before and after the incident. This is important information. Other specific questions include "What were you doing?" "How far from the incident were you?" "Do you wear glasses, and did you have them on?"

Other related information that may not be obvious includes on-the-floor "standard" process deviations, planned or unplanned plant shutdowns, time of year, weather conditions, geography (where did the failure happen?), and personnel losses. In fact, a good question to request on the eyewitness statement form is, "Are you aware of any recent changes in operations, personnel, equipment, etc.?" Will anyone voluntarily tell you about these things? No. Does it show up in any report? No. Does it show up in any paperwork? No. This kind of information is not apparent, so ask.

The information gathered so far comprises the basic facts concerning the failure. Now ask the question, "What do you think happened?" Most people want to tell you what they think happened. They have an idea. And anytime there is an incident in a plant, talk and theories will be floating around. Eyewitnesses want to tell you their opinions, especially after you've asked all the other questions. Compare their opinions to the facts, and see how they compare to one another.

I was involved with the failure of a very large 13-8Mo PH martensitic stainless steel forging that was being forged and heat treated in the East. The failures were sporadic and were discovered during acceptance testing due to the lack of mechanical properties. The physical cause of the failure was the creation, during heat treatment, of a sigma phase, which is a very brittle phase oriented in the transverse direction. When subjected to transverse direction tensile testing, the samples had absolutely zero ductility. It turned out that the forging company had no trouble cooling the large forgings to below -50 °C (-60 °F) during the heat treatment quench cycle in the wintertime when there was an abundance of snow outside to dump into the quench pit. During the hot summer months, however, the engineers had to calculate how much ice to have ready to dump into the quench tank. It became apparent that their calculations were not correct 100% of the time. This is a perfect example of how geographic location and time of year can be important to determining a failure's root cause.

Freeze the Evidence

Be consistent during each failure investigation. This approach will pay off. If you always request complete failure scene photographs, eyewitness statements, and all available information, your customers will start to collect everything before you are contacted. You will have trained them. If you work in a large company, it is a good idea to create a "What to do when a failure occurs" procedure and then train company personnel on how to properly document and protect the failure scene and how to properly fill out eyewitness statements. As a start, ASTM Committee E-30 on Forensic Sciences and ASTM Subcommittee E-30-05 on Forensic Engineering Sciences have created specifications that you can review and use when relevant. Some of these specifications are:

- E 620: Standard Practice for Reporting Opinions of Technical Experts
- E 678: Standard Practice for Evaluation of Technical Data
- E 860: Standard Practice for Examining and Testing Items That Are or May Become Involved in Litigation
- E 1020: Standard Practice for Reporting Incidents
- E 1188: Standard Practice for Collection and Preservation of Information and Physical Items by a Technical Investigator

Another source of information is the Federal Rules of Evidence, part of the Code of Federal Regulations, used for legal proceedings. An engineer involved in failures that will end up in legal proceedings must be aware of federal and local rules concerning evidence and testing. For example, it is important that failure investigators understand the legal rules concerning digital photographs. Are digital photographs unacceptable because they are too easy to alter? Many digital cameras will autocorrect a photograph. That is why having a known color standard in the photograph is a must. In fact, some investigators who specialize in corrosion failure still use film because it is more true to the color of the corrosion. It is also important to understand how to document transfer of evidence for purposes of testing. This is not an issue when a test is performed by an in-house lab, but "chain of custody" procedures must be followed when sending evidence to outside testing labs.

A good approach to "freezing" the evidence comes from a friend of mine in Canada. He altered a concept from the Failsafe Network, Inc. The following five P's must be documented and recorded in order to freeze evidence at the failure scene:

- *Position:* Fragments, equipment, parts, people (witnesses, personnel involved), controls, photographs. Not only is it good that you take the photographs, but have a layout of where you stood when you took them. Get a print of the building and document where you took which pictures.
- *People:* Job descriptions, witnesses, accountabilities, information sources, experts. Separate those personnel who know and do not know pertinent information.
- *Paper:* Drawings, design changes, manufacturing records, mill certificates, operation records, operating procedures, past failure histories,

maintenance records, photographs, inspection records, stress analysis, past nonconformances. Sometimes this information can be found, sometimes not; sometimes the records will be gone. Try to obtain all these pieces of information. Figure out whether the same failure has happened before and whether a failure analysis was performed or whether they just wrote up a nonconformance report, bought off the part, and moved on.

- *Process:* Design process, operational process, approved process changes, unapproved process changes (how it really operated or was made), environment, weather. There generally is a difference between what a company says it does and what actually happens on the shop floor. What happens on the job traveler is not exactly what happens in the shop. You need to observe the process and ask questions. "Do you guys follow this procedure exactly to the letter? Is this exactly the way you do the work?" If you are talking to the manager, the answer is, "Of course." If you are talking to the shop floor guy, the answer may be, "Well, we take this little thing and then we do it a little differently."
- *Parts:* Materials specified, materials used, mechanical and physical properties of materials for hardware or machine, fracture faces, distortion of failed part and other hardware, remnants of failed hardware or machine, microscopy analysis, stress analysis, metallurgical analysis.

Improper material selection rarely happens anymore. However, the one thing I do during every failure investigation is to check the material. It is an easy test to perform, so always do it. Every once in a while you will discover that it is the wrong stuff. It is rare, but it happens. If a material has been tensile tested, I ask for the tensile bars back so that I can examine them while looking over the test data. I have seen tensile test results that showed adequate mechanical properties. Unfortunately, the tensile bar was bent, indicating a bad test and that the data generated were incorrect to the low side due to complex loading. Such data cannot be trusted and should not be used for analysis of hardware. Hardware commonly is accepted based on tensile properties only. A part may fail because the mechanical properties actually were incorrect. In the end, the root cause of the failure is a human root cause. The test lab used by the company does not know how to perform a tensile test. And that could have been caught long before the part was made.

Step 3: Objectively and Clearly Identify All Possible Root Causes

The next step in the organization of a failure investigation is to *objectively* and *clearly* identify *all* possible root causes. Note the three very descriptive words used in the preceding sentence. This step means avoid chasing a single silver bullet theory. You must look at all possible root causes even when your manager or customer tells you, "This is what happened, now go prove it for me." In the end, he will not be around when the failure investigation becomes disorganized and falls to pieces.

Divide and Conquer. Many tools are at your disposal. One common tool used in failure investigations is a fault tree, an example of which is shown in Fig. 3. This organization tool is commonly used in systems engineering. Computers simplify creation and revision of fault trees and other analytical evaluation tools. The fault tree in Fig. 3 was created in Microsoft Word using the organization chart symbols. As shown in the fault tree, each root cause identified by brainstorming is given a number, and the related root causes under the same major category or grouping are given related numbers. This numbering system is important to maintaining organization during the failure investigation and will be carried forward on future documentation. The use of color can be useful for indicating the status of investigation results. For example, red could indicate a possible root cause that has been eliminated and green could indicate a root cause that has been proved.

The simplicity of the fault tree is that it divides a very complex problem or issue into more manageable components. The Romans conquered the known world with a very simple military maneuver that would divide

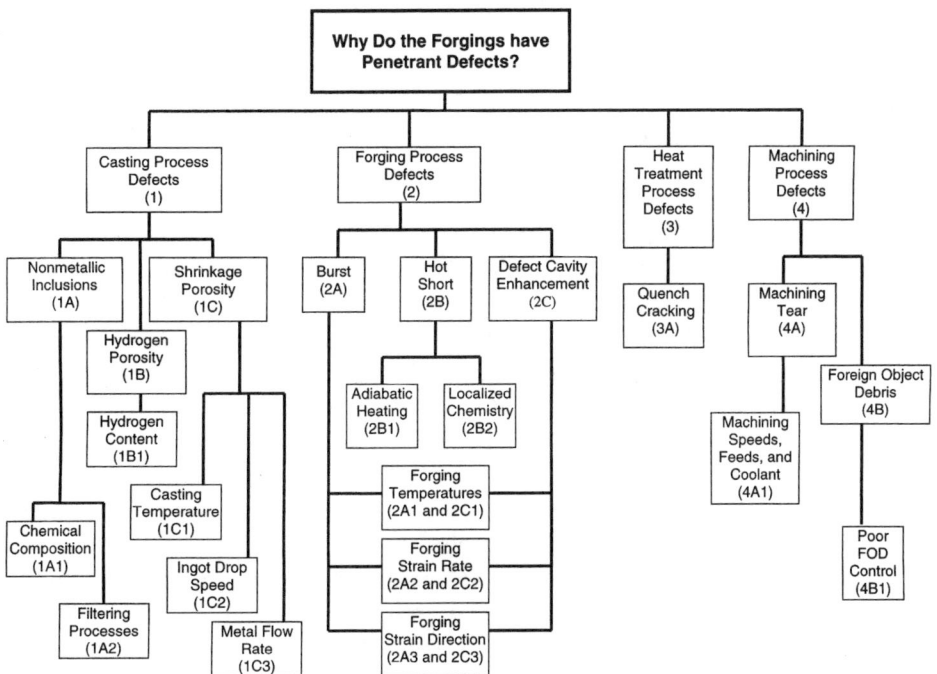

Fig. 3 Fault tree

their opponent's army into smaller units and make them easier to defeat. The Romans would march forward in large square blocks of soldiers, but the blocks were staggered. As long as the Roman blocks remained intact, the opposing forces would find themselves trapped on three sides by Roman soldiers. So, the next time you want to "defeat" a failure investigation, remember the Romans and the concept to "divide and conquer."

In fact, the whole concept behind a fault tree is "divide and conquer." By dividing a large failure investigation into smaller and more organized segments, you can attack each segment individually. The failure investigation team can better understand smaller problems and better handle the paperwork to solve them. Can you imagine the fault tree for the *Columbia* Space Shuttle accident in 2003 or the TWA plane crash in 1996? Absolutely huge. Attempting to fully comprehend all potential root causes of the failure and to keep them all straight would be mind boggling. Remember, divide and conquer.

The first few minutes of the movie *Gladiator* provide a good example of the Roman military maneuver to divide and conquer the Germanic tribes. The movie also dramatically demonstrates another common failure investigation event. There is a moment when the Germanic tribes realize they have forgotten about the Roman cavalry that is attacking from the right flank. The looks on their faces say it all: "We forgot about those guys." That same event happens when a failure investigation chases a single silver bullet theory and then realizes that the theory can't be proved. The engineer in charge of the investigation often has the same look on his face: "What do I do now?"

Brainstorming, as always, is a good tool. To discover every root cause I suggest you play the "Why, Why, Why" game. First, write down the failure in the form of a question beginning with the word "Why." This question is at the top of the fault tree (Fig. 3). Write down the first root cause answer and then ask why that first root cause answer happened. Continue the "Why, Why, Why" game until each line of questioning is exhausted or you have reached a practical stopping point. But do not stop too soon. For example: Why did this part fail? The material is bad. Why is the material bad? The material does not satisfy specification requirements. Why does the material not satisfy specification requirements? The chromium content is wrong. Why is the chromium content wrong? Keep asking why until you have exhausted the questions and answers. Then go back to the top of the fault tree and start again. In the example given here, stay in the bad material root cause section until all possible root causes for the material to be bad have been noted. There are multiple reasons why the material could be bad. The company bought the wrong material. The material was heat treated incorrectly. The material was contaminated during storage. Keep doing this until there are no more root cause answers to the "why" question. Be objective and list all answers; that is the key to brainstorming. If someone says that gremlins did it, write it down. The chance to be subjective comes later. It is not appropriate to dismiss any idea. The dismissal violates the brainstorming concept, impedes the flow of ideas from other participants, and possibly ends the brainstorming session.

Examine the fault tree shown in Fig. 3. Why do the forgings have penetrant defects? When you attempt the "Why, Why, Why" game, you realize you cannot play until you have collected as much information concerning the failure as possible. You have to know that there was a casting process, a forging process, and a heat treatment process in the fabrication sequence. You have to have gathered up enough information about how the part was fabricated in order to play the game. In the "Step 2" section of this chapter, we noted that the failure investigator must become an expert about the failed part and its fabrication. You must gather enough information to become this expert, or enough information to know which personnel are required to support the failure investigation (e.g., experts from the raw material supplier, the forging supplier, the heat treat supplier, the machine shop, and the testing laboratory). If you do not have the expertise yourself, it is imperative to have knowledgeable people in the room during the brainstorming session. If the failure investigation team does not have the correct expertise, then the fault tree will have missing pieces.

Once the fault tree is completed, two other questions should be asked: "What was different about this failure?" and "What are we missing?" The first question requires the team to focus on the differences between the failed component or process and the successful component or process, instead of the similarities.

The second question is a sanity check to make sure the team continues to think of reasons the failure occurred rather than focusing exclusively on the reasons already presented. You never want to look up and see the Roman cavalry coming from the right flank because you forgot to consider something. During the course of the failure investigation, pause, take a big step back, and ask, "Are we missing something here?" This is more difficult than it sounds. Once the failure investigation starts, day-to-day activities and monitoring progress will become all-consuming. Another approach is to ask someone else if they see anything missing. That person may not be so involved and can look at the failure investigation with a more unbiased viewpoint. Finally, during brainstorming there are people who love to speak up and people who do not like to speak up in a crowd. Always end your brainstorming session with the opportunity for people to talk to you later one on one about anything they would like to add. Those quieter people will show up.

As the failure investigation proceeds, new information will surface and a new review of the data will be required. Ask these same two questions over and over, throughout the failure investigation.

Importance of a Fault Tree. First and foremost, a fault tree provides documentation. It is a permanent record of all possible root causes imagined and developed by the team. No ideas are lost or forgotten. In addition, this chart is excellent for briefings. All root causes are succinctly listed and organized. It is a very good presentation tool for instilling confidence in management and your customer that the failure investigation is under control.

Second, the fault tree is a "living" list that can be added to at any time. Do not stick it in a drawer once the failure investigation begins. It will improve and change over time. As the team learns something new during the natural progression of the investigation, add the information to the fault tree. Information will be generated that will prove or disprove each root cause on the fault tree and that may also lead to a new root cause not imagined before.

Third, the fault tree helps to simplify a very complicated process. As noted earlier, the Romans conquered the known world by using military maneuvers to divide and conquer their foes. The fault tree is based on the same concept. The failure is divided into individual root causes, each of which is further divided. It is easier to investigate and prove or disprove each root cause than tackle the entire resultant failure.

Fourth, the fault tree should help prevent the dreaded "silver bullet" approach to failure investigation. There is always a predominant or "pet" root cause. Resist the urge to spend all the time, money, and labor to prove this one root cause. If wrong, you have wasted resources and perhaps destroyed any evidence that would prove or disprove the other root causes. Not all failure investigations have enough time, personnel, and available test samples to test every root cause. You may have to be selective in your approach. Besides, the failure may be due to a combination of root causes, not just one. Just because you prove the "pet" root cause does not mean there is not another, more important, root cause out there. Remember that the goal of a failure investigation is to determine the root cause(s) of a failure, not to prove one root cause. Sometimes the failure investigation goes off track and becomes nothing more than a proof of a single theory.

Finally, the fault tree can be altered to show "and/or" conditions. Sometimes root causes have a cause-and-effect relationship. One root cause cannot occur without another root cause occurring simultaneously or previously. There may also be a synergistic effect between root causes. The fault tree can present both of these relationships. This is good information, because if you prove that one of the root causes cannot occur, you have automatically ruled out the related root cause(s).

As the failure investigation progresses, various root causes can be crossed out to indicate their elimination. Add the elimination date for each discarded root cause. For a more colorful chart, use green type for the open root causes and red for the eliminated ones.

The fault tree is the first step in failure investigation organization.

Step 4: Objectively Evaluate the Likelihood of Each Root Cause

Prioritize the Root Causes. The next step of the failure investigation is to objectively evaluate the likelihood of each root cause listed in the fault tree. A good tool for this is the failure mode assessment (FMA) chart. The FMA chart shown in Fig. 4 was created in a Microsoft Excel spreadsheet and comes straight from the fault tree in Fig. 3. Note how the numbering systems on the fault tree and the FMA chart are identical and easy to follow.

The fault tree has provided all of the root causes, but where do you start the investigation? The FMA chart is useful in deciding a course of action.

As noted previously, not all failure investigations have enough time, personnel, and available test samples to test every root cause. You may have to be selective in your approach. First, assess the probability of each potential root cause. How realistic is it? Assign each root cause a probability, such as likely, possible, or unlikely. The best approach is to use the technical expertise of the failure investigation team. By being subjective and selective, the team will determine which root cause or causes it will attempt to prove or disprove first.

Second, assign a priority to each potential root cause. In what order will the investigation evaluate each root cause? Factors that affect this decision are the same as noted previously: hardware availability for testing, costs, and time. In addition, consider the effect of the planned test on other tests. The testing sequence must ensure that one test will not eliminate or contaminate a later test. Also, perhaps one test can be used to disprove or prove multiple root causes. These considerations will help to prioritize the possible root causes and therefore the order of testing.

Third, use the FMA chart to document the rationale used to assign a root cause its particular probability and priority. Once the failure investigation starts, it may be difficult to remember why a certain test was performed first or last. Lastly, any action items directed toward proving or disproving the root cause should also be documented in an abbreviated manner.

Importance of an FMA Chart. The FMA chart is a permanent record of the assessed probability and assigned priority of each root cause, the rationale for the particular assessments, and the action items to prove or disprove each root cause. Once again, this chart is an excellent tool for briefings. The failure investigation is now documented on two easy-to-understand charts: the fault tree and the FMA.

Like a fault tree, an FMA chart is a "living" list that can be changed at any time. Once the failure investigation is in progress, information will be generated that will prove or disprove each root cause on the fault tree. This information may also lead to a new probability or priority rating. A

No.#	Potential Root Cause	Probability	Priority	Rationale	Technical Plan for Resolution
1	**Casting Process**				
1A	**Non-Metallic Inclusions**	Likely	1	I) Failure Analysis of S/N 1 discovered Al Oxide inclusion in Defect Cavity by Metallographic exam. Inclusion confirmed to be Al oxide with presence of Silicon by SEM & EDS Analysis. II) Grain flow around Defect Cavity discovered during failure analysis of S/N 1 & S/N 2 by Metallographic Exam.	1) Review of Casting Process with Both Suppliers. 2) Failure analysis of S/N 28 including metallographic examination, enhanced NDT evaluation and mechanical property testing. 3) Complete Forging History Spreadsheet for trend analysis and problem definition - Class AAA level UT inspection of forgings at Assembly Facility required.
1A1	Chemical Composition	Not likely	2	Chemical Compositions are within Spec	1) Check Chemical Composition meets AMS spec for each ingot. 2) Compare Chemical Compositions for any variations
1A2	Filtering Processes	Likely	1	Filtering Processes control non-metallic inclusion level in ingots	1) Review Supplier B Casting Process in general and the triple filter system in specific.
1B	**Hydrogen Porosity**	Not likely	2	I & II above. III) Hydrogen analysis of S/N 1&7 by Supplier A indicates low potential for hydrogen porosity. IV) This type of casting defect should heal during forging process.	1) Review of Casting Process with Both Suppliers. 2) Failure analysis of S/N 28 including metallographic examination, enhanced NDT evaluation and mechanical property testing. 3) Complete Forging History Spreadsheet for trend analysis and problem definition - Class AAA level UT inspection of forgings at Assembly Facility required.
1B1	Hydrogen Content	Not likely	2	I-IV above	1) Review Supplier B Casting Process in general and the hydrogen content testing in specific.
1C	**Shrinkage Porosity**	Not likely	2	I & II above. III) Pore count of S/N 1&7 by Supplier B indicates low potential for shrinkage porosity. IV) This type of casting defect should heal during forging process.	1) Review of Casting Process with Both Suppliers. 2) Failure analysis of S/N 28 including metallographic examination, enhanced NDT evaluation and mechanical property testing. 3) Complete Forging History Spreadsheet for trend analysis and problem definition - Class AAA level UT inspection of forgings at Assembly Facility required.
1C1	Casting Temperature	Not likely	2	I-III above	1) Review Supplier B Casting Process in general and casting procedures in specific.
1C2	Ingot Drop Speed	Not likely	2	I-III above	1) Review Supplier B Casting Process in general and casting procedures in specific.
1C3	Metal Flow Rate	Not likely	2	I-III above	1) Review Supplier B Casting Process in general and casting procedures in specific.
2	**Forging Process**				
2A	**Burst**	Not likely	2	I & II above. III) SEM analysis of S/N 1 & S/N 2 Defect Cavity surface not indicative of forging burst.	1) Review of Forging Process with Both Suppliers. 2) Failure analysis of S/N 28 including metallographic examination, enhanced NDT evaluation and mechanical property testing. 3) Complete Forging History Spreadsheet for trend analysis and problem definition - Class AAA level UT inspection of forgings at Assembly Facility required.
2A1	Forging Temperatures	Not likely	2	I-III above	1) Review Supplier A Forging Process in general and hot working procedures in specific
2A2	Forging Strain Rate	Not likely	2	I-III above	1) Review Supplier A Forging Process in general and hot working procedures in specific
2A3	Forging Strain Direction	Not likely	2	I-III above	1) Review Supplier A Forging Process in general and hot working procedures in specific
2B	**Hot Short (Adiabatic Heating)**	Not likely	2	I & II above. III) SEM analysis of 1 & S/N 2 Defect Cavity surface not indicative of hot tear.	1) Review of Forging Process with Both Suppliers. 2) Failure analysis of S/N 28 including metallographic examination, enhanced NDT evaluation and mechanical property testing. 3) Complete Forging History Spreadsheet for trend analysis and problem definition - Class AAA level UT inspection of forgings at Assembly Facility required.
2B1	Adiabatic Heating	Not likely	2	I-III above	1) Review Supplier A Forging Process in general and hot working procedures in specific
2B2	Localized Chemistry Variation	Not likely	2	I-III above	1) Review Supplier A Forging Process in general and hot working procedures in specific
2C	**Defect Cavity Enhancement**	Possible	1	I) S/N 1 Defect Cavity had no obvious inclusion and S/N 2 Defect Cavity was not filled with Al Oxide inclusion. II) Size of Defect Cavity unique.	1) Review of Forging Process with Both Suppliers. 2) Failure analysis of S/N 28 including metallographic examination, enhanced NDT evaluation and mechanical property testing. 3) Complete Forging History Spreadsheet for trend analysis and problem definition - Class AAA level UT inspection of forgings at Assembly Facility required.
2C1	Forging Temperatures	Possible	1	I-III above	1) Review Supplier A Forging Process in general and hot working procedures in specific
2C2	Forging Strain Rate	Possible	1	I-III above	1) Review Supplier A Forging Process in general and hot working procedures in specific
2C3	Forging Strain Direction	Possible	1	I-III above	1) Review Supplier A Forging Process in general and hot working procedures in specific
3	**Heat Treatment Process**				
3A	**Quench Cracking**	Not likely	3	Quench cracking does not normally produce small, multiple internal flaws.	Supplier A investigating cause of quench cracks in S/N 5 and S/N 6
4	**Machining Process**				
4A	**Machining Tear**	Not likely	3	No evidence of tearing, smearing, etc. on part surface or in Defect Cavity.	None
4A1	Machining Speeds, Feeds & Coolant	Not likely	3	No evidence of tearing, smearing, etc. on part surface or in Defect Cavity.	None
4B	**Foreign Object Debris**	Not likely	3	No FOD discovered on part surface or in Defect Cavity.	None
4B1	Poor FOD Control	Not likely	3	No FOD discovered on part surface or in Defect Cavity.	None

Fig. 4 Failure Mode Assessment (FMA) chart. Forging with defects

quick glance at the FMA chart will show what will happen if you change the probability or priority of one root cause. Once you have created the chart, *use* it.

The FMA chart provides the order in which each root cause will be proved or disproved. This provides more failure investigation organization and leads directly to the next step.

Step 5: Converge on the Most Likely Root Cause(s)

Devising a Technical Plan. Now it is time to converge on the most likely root cause(s). This correlates with step 2 in the four-step problem-solving process ("What is the root cause of the problem?") discussed in Chapter 3. A useful tool is the technical plan for resolution (TPR) chart. The TPR chart shown in Fig. 5 was created in a Microsoft Excel spreadsheet and comes straight from the FMA chart in Fig. 4. Again, note how the numbering system in Fig. 3 to 5 is identical and easy to follow. In fact, a good practice is to tie the FMA and TPR charts together on two sheets within one spreadsheet file so that any changes made to the FMA will be immediately reflected in the TPR.

The fault tree and FMA chart have provided the possible root causes, assessed the probability of each, and prioritized them. Now comes time to create the technical plan to prove or disprove each root cause. The TPR chart serves as a detailed road map, ensuring that the testing or analysis achieves maximum effect and efficiency. Once the details are down on paper, a specific order—or synergy—of testing or analysis will become evident. You may change the priority of a certain test because it shows up under multiple root causes and is a good test to perform early. Alternatively, you may discover that you will run out of material by the time you get to a certain test or analysis. Examples of the types of tests and analyses to perform can be found in Appendix 1, "General Procedures for Failure Analysis."

A good practice at this stage is to ask yourself and your team three questions about the failure:

- What do I know?
- What do I think?
- What can I prove?

First ask, "What do I know?" List the hard facts and existing data, and then determine to which root cause(s) they apply.

When the list of facts is complete, it's time to ask "What do I think?" Certain theories, ideas, and possible cause-and-effect relationships will arise from history, experience, logic, and the hard facts noted above. This stage is particularly enjoyable for team members who like to ask "What if?"

No.#	Potential Root Cause	Priority	Technical Approach for Resolution	Who?	When?	Result?
1	Casting Process					
1A	Non-Metallic Inclusions	1	1) Review of Casting Process with Both Suppliers.	1) Team	3/6/01	1) Casting Process, not alloy conclusion 2) Casting Station Checklist Requested - Received 3/16/01 But not used on our Ingots - Received Actual checklist on 3/23/01 / Much smaller & less defined 3) Checklist review requested - No violations from Supplier B Procedure found on all ingots 4) Mr. Smith Concerned with Lack of CFF inspections!!
		1	2) Failure analysis of S/N 28 including metallographic examination, enhanced NDT evaluation and mechanical property testing.	2) Engineering Center	Failure Analysis Phase 1 - 4/27/01 Failure Analysis Phase II - ???	
		1	3) Complete Forging History Spreadsheet for trend analysis and problem definition - Class AAA level UT inspection of forgings at Assembly Facility required.	3) Assembly Facility	10 per month - 7/1/01	
1A1	Chemical Composition	2	1) Check Chemical Composition meets AMS spec for each ingot.	1) Supplier B	4/1/01	
		2	2) Compare Chemical Compositions for any variations	2) Supplier B	4/1/01	
1A2	Filtering Processes	1	1) Review Supplier B Casting Process in general and the triple filter system in specific.	1) Team	3/6/01	1) Housekeeping Questions 2) Same as 1A 3) Same as 1A 4) Same as 1A
1B	Hydrogen Porosity	2	1) Review of Casting Process with Both Suppliers.	1) Team	3/6/01	1) Casting process & hydrogen checks ok 2) Same as 1A 3) Same as 1A 4) Same as 1A
		2	2) Failure analysis of S/N 28 including metallographic examination, enhanced NDT evaluation and mechanical property testing.	2) Engineering Center	Failure Analysis Phase 1 - 4/27/01 Failure Analysis Phase II - ???	
		2	3) Complete Forging History Spreadsheet for trend analysis and problem definition - Class AAA level UT inspection of forgings at Assembly Facility required.	3) Engineering Center	In Work - Constantly updated	No trends established so far
1B1	Hydrogen Content	2	1) Review Supplier B Casting Process in general and the hydrogen content testing in specific.	1) Team	2/28/01	1) Hydrogen content low 2) Same as 1A 3) Same as 1A 4) Same as 1A
1C	Shrinkage Porosity	2	1) Review of Casting Process with Both Suppliers.	1) Team	3/6/01	1) Casting process acceptable 2) Same as 1A 3) Same as 1A 4) Same as 1A
		2	2) Failure analysis of S/N 28 including metallographic examination, enhanced NDT evaluation and mechanical property testing.	2) Engineering Center	Failure Analysis Phase 1 - 5/9/01 Failure Analysis Phase II - ???	
		2	3) Complete Forging History Spreadsheet for trend analysis and problem definition - Class AAA level UT inspection of forgings at Assembly Facility required.	3) Engineering Center	In Work - Constantly updated	No trends established so far
1C1	Casting Temperature	2	1) Review Supplier B Casting Process in general and casting procedures in specific.	1) Team	3/6/01	Same as 1C
1C2	Ingot Drop Speed	2	1) Review Supplier B Casting Process in general and casting procedures in specific.	1) Team	3/6/01	Same as 1C
1C3	Metal Flow Rate	2	1) Review Supplier B Casting Process in general and casting procedures in specific.	1) Team	3/6/01	Same as 1C
2	Forging Process					
2A	Burst	2	1) Review of Forging Process with Both Suppliers.	1) Team	3/7/01	1) History of defects in rings. Much smaller percentage than current alloy. 2) Two Forgings with quench cracks in review. 3) Forging Temperature increase due to Adiabatic Heating not expected but not checked. 4) Forging procedure review requested - Completed 3/26/01 - No Deviations from Supplier A Procedure found for all forgings
		2	2) Failure analysis of S/N 28 including metallographic examination, enhanced NDT evaluation and mechanical property testing.	2) Engineering Center	Failure Analysis Phase 1 - 4/27/01 Failure Analysis Phase II - ???	
		2	3) Complete Forging History Spreadsheet for trend analysis and problem definition - Class AAA level UT inspection of forgings at Assembly Facility required.	3) Engineering Center	In Work - Constantly updated	No trends established so far
2A1	Forging Temperatures	2	1) Review Supplier A Forging Process in general and hot working procedures in specific	1) Team	3/7/01	Same as 2A
2A2	Forging Strain Rate	2	1) Review Supplier A Forging Process in general and hot working procedures in specific	1) Team	3/7/01	Same as 2A
2A3	Forging Strain Direction	2	1) Review Supplier A Forging Process in general and hot working procedures in specific	1) Team	3/7/01	Same as 2A
2B	Hot Short (Adiabatic Heating)	2	1) Review of Forging Process with Both Suppliers.	1) Team	3/7/01	Same as 2A
		2	2) Failure analysis of S/N 28 including metallographic examination, enhanced NDT evaluation and mechanical property testing.	2) Engineering Center	Failure Analysis Phase 1 - 4/27/01 Failure Analysis Phase II - ???	
		2	3) Complete Forging History Spreadsheet for trend analysis and problem definition - Class AAA level UT inspection of forgings at Assembly Facility required.	3) Engineering Center	In Work - Constantly updated	No trends established so far
2B1	Adiabatic Heating	2	1) Review Supplier A Forging Process in general and hot working procedures in specific	1) Team	3/7/01	Same as 2B
2B2	Localized Chemistry Variation	2	1) Review Supplier A Forging Process in general and hot working procedures in specific	1) Team	3/7/01	Same as 2B
2C	Defect Cavity Enhancement	1	1) Review of Forging Process with Both Suppliers.	1) Team	3/7/01	Same as 2B
		1	2) Failure analysis of S/N 28 including metallographic examination, enhanced NDT evaluation and mechanical property testing.	2) Engineering Center	Failure Analysis Phase 1 - 4/27/01 Failure Analysis Phase II - ???	
		1	3) Complete Forging History Spreadsheet for trend analysis and problem definition - Class AAA level UT inspection of forgings at Assembly Facility required.	3) Engineering Center	In Work - Constantly updated	No trends established so far
		1	4) Search of NR tags for other Supplier A forgings from other alloys	4) Engineering Center	2/21/01	NR tag information inadequate - cannot discern type of defect, forging ID, etc.
2C1	Forging Temperatures	1	1) Review Supplier A Forging Process in general and hot working procedures in specific	1) Team	3/7/01	Same as 2C
2C2	Forging Strain Rate	1	1) Review Supplier A Forging Process in general and hot working procedures in specific	1) Team	3/7/01	Same as 2C
2C3	Forging Strain Direction	1	1) Review Supplier A Forging Process in general and hot working procedures in specific	1) Team	3/7/01	Same as 2C
3	Heat Treatment Process					
3A	Quench Cracking	3	1) Supplier A investigating cause of quench cracks in S/N 5 and S/N 6	1) Supplier A	4/1/01	
4	Machining Process					
4A	Machining Tear	3	None	N/A	N/A	N/A
4A1	Machining Speeds, Feeds & Coolant	3	None	N/A	N/A	N/A
4B	Foreign Object Debris	3	None	N/A	N/A	N/A
4B1	Poor FOD Control	3	None	N/A	N/A	N/A

Fig. 5 Technical Plan for Resolution (TPR) chart. Forging with defects

Lastly, ask the question, "What can I prove?" For each of the theories, ideas, or root causes, list the physical evidence you would expect to see if that a particular root cause occurred. Then list the tests and analyses required to discover that physical evidence. These will prove or disprove each root cause. It is sometimes easier to disprove a theory for the root cause than to prove it. This may involve testing or analysis of the failed hardware or similar hardware, exemplar testing to demonstrate if a failure mode is possible, operational data analysis, stress analysis, literature searches, phone calls to other divisions or companies that may have encountered a similar problem, and so on.

For example, when the root cause is hydrogen embrittlement of high-strength steels, some of the expected physical evidence in the metallographic sample would be intergranular cracking and a fracture surface with a rock-candy appearance. The failure investigation should also discover an environmental condition or fabrication process that generates hydrogen and a tensile stress of some kind.

The answers to the "What can I prove?" question create the TPR chart. This chart is where you plan all the testing and analysis required to prove or disprove the root causes.

This is also where the failure investigation team must make the difficult decisions regarding the sequence of testing and analysis: Is there enough material, enough time, enough personnel and money? Using the assessed probability and assigned priority of each root cause in the FMA chart, the team must decide which root causes to focus on first and thus which tests should be performed first. The team also must decide whether to work in parallel or in series. The goal is a failure investigation performed in the most organized and efficient manner possible with systematic and coordinated testing. Sometimes the result of one test will eliminate the need for another test because the result provides new or unexpected information. For example, if you perform a tensile test on the material and the results are satisfactory, then the failure investigation progresses in one direction. If the results are unsatisfactory, however, then the investigation progresses in another direction. Or does it? What do you do next?

Remember that, due to time and/or other constraints, you may not be able to evaluate every root cause. Therefore, you must be organized, efficient, and smart when planning the work to be performed. When you lay out the TPR chart, the order of testing will be apparent.

The TPR chart is the failure investigation plan. It is the detailed test and analysis road map for proving or disproving each root cause. The TPR addresses each test, examination, or record review, letting you know whether you need to perform a tensile test or a stress corrosion test, check certain records, or visit the supplier and look for details A, B, and C.

The TPR chart also documents three additional factors: (1) the responsible party for getting a test or analysis performed, (2) the completion date, and (3) the results. I cannot stress highly enough that the person assigned must agree both to perform the test or analysis and to meet the completion date. Merely assigning a task does not always ensure that it gets done, but if someone agrees to the task and the established deadline, then the work normally is accomplished. Record the test results on the chart, along with any related comments.

Importance of a TPR Chart. The TPR chart is a permanent record of each test or analysis that is to be performed, the person assigned, the expected completion date, and the results. You now have the failure investigation documented on three easy-to-understand charts: the fault tree, the FMA, and the TPR.

Like the other two charts, the TPR is a "living" document and can be changed at any time. Once the failure investigation is in progress, information will be generated that will prove or disprove each root cause on the fault tree. This information may also lead to new root causes or to new probability or priority ratings. When these changes occur on the fault tree or FMA, the TPR chart will also change. However, a quick glance at the TPR chart will inform you whether the new change will require more testing or if the required testing is already planned. The three charts are easy to read because they are tied together with the same number and root cause set up in the fault tree and the same probability and priority set up in the FMA chart.

Congratulations, you now have an organized failure investigation! The three charts described in the preceding sections provide a living documentation of the failure analysis. These "Big Three" charts can be pulled up on a moment's notice and presented to document the status of any project. There is not a question you cannot answer. There is not a who, what, when, where, and rationale that is not readily apparent. With these three charts you can handle any failure investigation of any size.

Step 6: Objectively and Clearly Identify All Possible Corrective Actions

After weeks or months of hard work, the organized and efficient failure investigation planned out above will determine the root cause(s). The failure investigation will have answered two questions of the four-step problem-solving process discussed in Chapter 3: "What is the problem?" and "What is the root cause of the problem?" However, the failure investigation is not yet complete. In fact, you are only halfway finished. Unfortunately, many failure investigations end here, but now you need to determine the corrective action(s) to prevent the failure from occurring again.

The next step is to identify all possible corrective actions. This is step 3 in the four-step problem-solving process discussed in Chapter 3.

If step 6 sounds familiar, it should. This step is essentially a repeat of step 3 in the nine-step process outlined in the present chapter: "Objectively and clearly identify all possible root causes." The difference is that the failure investigation team is now focused on identifying all possible corrective actions to solve the root cause(s) identified in the failure analysis. In reality, each corrective action usually becomes apparent as each root cause is identified during the failure investigation. Repeat the instructions in step 3 and create a corrective action tree. The corrective action tree shown in Fig. 6 was created in Microsoft Word using organization chart symbols.

This chart can be used for preventive action as well as corrective action. Corrective action is something you must do to ensure the root cause of the failure will not happen again. Preventive action involves taking steps to discover the root cause, such as improved monitoring.

Figure 6 indicates various potential corrective actions for the failure involving forgings with penetrant defects. A brainstorming effort produced all the corrective actions noted on the chart. The major corrective actions include the casting process, the forging process, the heat treatment process, and nondestructive evaluations. Three of the corrective actions are directly related to the fault tree (Fig. 3), but the nondestructive evaluations constitute a new category for preventive action. The four major

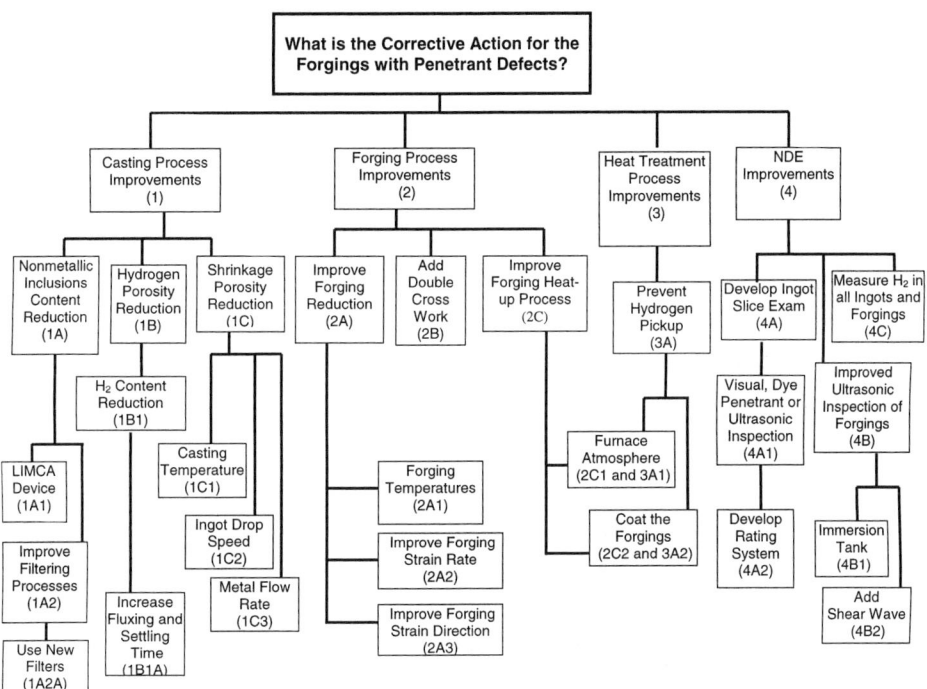

Fig. 6 Corrective action tree

corrective actions are numbered, as are the minor corrective actions related to each.

Step 7: Objectively Evaluate Each Corrective Action

Step 7 involves objectively evaluating the likelihood of each corrective action listed in the corrective action tree. This step is essentially a repeat of step 4, "Objectively evaluate the likelihood of each root cause." The difference is that the failure investigation team is assessing the probability and assigning priority to each corrective action, combined with a rationale. Repeat the instructions given in step 4 and create a corrective action assessment (CAA) chart.

Referring to the corrective action tree shown in Fig. 6, the best approach would be to incorporate all the corrective actions on the CAA chart. However, this may not be technically feasible, cost effective, or timely. The CAA chart should assess the probability of each corrective action to resolve the root cause or causes determined in the failure investigation. The technical experts should then assign a priority as to which corrective action will be evaluated for effectiveness. Once again, you are trying to figure out where you are going to spend your time, labor, and money in order to select the best corrective action. It may be unrealistic to think you can evaluate every corrective action. Which corrective action provides the most bang for your buck? In addition, this step must distinguish between corrective actions that eliminate the root cause and corrective actions that are preventive action to discover or monitor the root cause.

Step 8: Select the Optimal Corrective Action(s)

Now it is time to select the optimal corrective action(s). This relates to step 4 in the four-step problem-solving process, and is essentially a repeat of step 5 in the nine-step process: "Converge on the most likely root cause(s)." In this step the failure analysis team creates a technical plan to determine the best corrective action(s). Repeat the instructions given in step 5 and create a technical plan for evaluation (TPE) chart.

Step 9: Evaluate the Effectiveness of the Selected Corrective Action(s)

The last step is to recommend an evaluation of the corrective action(s) selected. After some time period, the corrective action should be evaluated. Whether the evaluation should be performed once, or periodically, is at the discretion of the failure investigation team. It is important that the evaluation be a proper test of the corrective action.

Case History: Rocket Fire, Former Soviet Union (1960)

Leonard C. Bruno

Unexpected ignition of the fourth stage section of a rocket undergoing repairs after a failed launch attempt killed a Soviet Field Marshal.

Background

On October 24, 1960, Field Marshall Mitrofan Nedelin and scores of Soviet space engineers and technicians were killed when a four-stage launch vehicle undergoing repairs on the launch pad burst into flames. The high death toll resulted from Nedelin's decision to attempt repairs immediately after a failed launch attempt, bypassing the basic safety measures of draining the rocket's fuel tanks and deactivating its electrical systems. When the launch vehicle's fourth stage ignited its engines, the space probe was turned into a mammoth fire bomb. The Soviet Union largely withheld information about the disaster; two decades passed before details become available to the West.

During the early years of the space age, when the Cold War between the United States and the former Soviet Union spurred the development of rival space programs, the Soviets hoped to set a record for sending the first space probe (unmanned satellite) to Mars and Venus. Such space travel, in additional to the technological requirements, is subject to calendar-based constraints, since the logistics of an interplanetary voyage are controlled by orbital paths. In the case of Mars, a "launch window" opens up for a few days every 26 months, allowing a satellite to be launched from Earth with a minimum expenditure of propulsion power and fuel. A similar window opens to Venus every 18 months.

During the September 1960 visit of Soviet Premier Nikita S. Khrushchev to the United Nations headquarters in New York, Khrushchev hinted and bragged that such an interplanetary launch was about to take place. There were even rumors that he had brought along souvenir models of the Soviet interplanetary probes to give away during his visit. No such launch occurred during Khrushchev's trip to the United Nations, however, and his return home was followed by the report that Soviet Field Marshal Nedelin had been killed in an airplane crash. At this time, there were unsubstantiated reports that a manned launch attempt had somehow ended disastrously. Among the stories circulating was a December report from an Italian news agency claiming that a space-vehicle explosion had killed not only Nedelin, but 100 Soviet scientists, engineers, and workmen as well.

Allegations that such a disaster had occurred did not surface again until the 1965 publication of *The Penkovskiy Papers,* the memoirs of a Soviet official who was executed for passing state secrets to the West. Penkovskiy provided few details about the accident, but did indicate that Nedelin and over 300 people were killed when they approached the rocket pad 15 or 20 m after the vehicle had failed to launch. The next mention of the disaster also appeared in Khrushchev's posthumously published memoirs. Published in 1974, these papers essentially duplicated Penkovskiy's account. In 1976, the Soviet exile Zhores Medvedev confirmed the accounts of Penkovskiy and Khrushchev by describing the event as "an irreparable castastrophe."

It was not until 1982, however, that the Soviet émigré S. Tiktin provided what many now consider the most plausible and reliable account of the event. Tiktin was responding in writing to the speculations of James E. Oberg, an American expert on the Soviet space program, about the disaster.

Details of the Fire

The Soviets were using the newly developed SS-7 ICBM (intercontinental ballistic missile) for the space probe that Khrushchev bragged about during his September 1960 trip to New York. Bugs had not entirely been eliminated from the system, however, and the politically charged project was subject to several delays. The launch was finally scheduled for October 24, 1960, under the supervision of Field Marshal Nedelin. When the countdown for the launch expired and the rockets failed to ignite, Nedelin must have felt compelled to go to any lengths to avoid yet another major delay.

Once the launch had clearly failed, Nedelin emerged from the protective bunker and indicated that the necessary repairs to the fuel-feed system would be made then and there, without performing the time-consuming safety precaution of draining the fuel tanks. This safety principle had been in force ever since the Germans began their liquid-fuel rocket experiments during World War II, and only someone of Nedelin's stature in the Soviet bureaucracy could have initiated such an order.

A similar gamble had paid off in July 1960, during the launch of an unmanned Vostok spacecraft. In this case, workers were able to weld a leak shut in the booster (first-stage) rocket while the launch vehicle was still on the pad and without draining out all of its dangerous propellants. Nedelin may have had this earlier, successful gamble in mind when he ordered his crew to proceed immediately with repairs. Furthermore, as Nedelin undoubtedly recognized, his willingness to oversee the repairs personally would have a reassuring effect.

Tiktin's report of the disaster explains that Nedelin and his crews set to work without realizing that the launch vehicle's fourth stage had for some reason not received the stand-down (temporary shut-off) command. According to the electronic information processed by the fourth stage, the launch had indeed occurred as planned. Thus, as the work supervised by Nedelin proceeded, the timer that controlled the fourth stage was ticking away, just as if the rockets of the first and second stages had completed their work and then fallen away from the upward-streaking rocket. Within a short period of time, this timer indicated that the small third stage was firing and placing the rocket into a low parking orbit. Following this stage, the fourth stage would ignite its 2000 pounds of fuel to escape from the earth's gravitational pull and shoot on toward Mars.

When the time came for the fourth stage to fire its rockets, it was in fact sitting on top of an inert but fully fueled launch vehicle. The fourth stage was 37 m (120 ft) above the launch pad, and only inches from the top of the third stage's fuel tank. When the fourth stage fired its engines, it took only seconds for the flames to burn through the thin metal top of the third stage's fuel tank and detonate the 18 metric tons (20 tons) of kerosene contained there. The 450 metric tons (1 million pounds) of kerosene and liquid oxygen held in the two large, lower stages would then have ignited immediately.

Tiktin described the ensuing explosion as a roaring, monstrous fire rather than a destructive concussion. Those individuals unfortunate enough to be on the pad at the time of the explosion were instantly immolated, while those inside the bunker were unharmed. Remarkably, some of the bus drivers who had brought the crew to the pad and were sitting at the wheel 90–180 m (100–200 yards) away from the explosion received only slight burns from scattering kerosene. Bus windows were not even shattered.

Impact

The only official Soviet reference to the accident was the brief announcement that Nedelin had died on October 25, 1960 in a plane crash. Nearly 20 years later, an official biography of Nedelin made no mention of an airplane crash as the cause of his death, stating only that Nedelin died "in the performance of his official duties." Following Tiktin's 1982 account of the accident, the Soviet government neither confirmed nor denied the story.

Tiktin's attempt to explain how such a seasoned space professional as Nedelin could violate the most basic principle of rocket safety places the blame on politics. Although Nedelin would have been fully aware of the risk he was taking, he responded to the enormous political pressure upon him to produce a successful launch to Mars. He was simultaneously constrained by the inflexible "launch window" inherent in every interplanetary attempt. Evidently, Nedelin felt he had no choice but to gamble.

This disaster cost the Soviet Union not only the experience and leadership of Mitrofan Nedelin, but a substantial portion of its skilled launch vehicle technicians. The loss of this body of expertise represented a major set-back to the Soviet's manned space program. If the explosion had not occurred, the epoch-making space flight manned by Uri Gagarin might have taken place even earlier than April 12, 1961. But perhaps the ultimate lesson of this catastrophe, which took the lives of anywhere from 40 to 300 individuals, is the often unforgiving nature of space technology—especially when politics takes precedence over safety.

SELECTED REFERENCES

- Z. Medvedev, "Two Decades of Dissidence," *New Scientist,* November 4, 1976, p 264–267
- D. Newkirk, *Almanac of Soviet Manned Space Flight,* Gulf Publishing, 1990
- J.E. Oberg, *Red Star in Orbit,* Random House, 1981
- J.E. Oberg, *Uncovering Soviet Disasters: Exploring the Limits of Glasnost,* Random House, 1988
- O. Penkovskiy, *The Penkovskiy Papers,* Avon, 1965
- S. Talbott, Ed., *Khrushchev Remembers: The Last Testament,* Little, Brown, 1974
- S. Tiktin, "U.S.S.R. and Space," *Posev,* October 1982, p 46–50

Case History Discussion

During the space race of the 1960s when the Cold War engineered rival space programs, many accidents occurred that were not reported or confirmed for almost 20 years. The sad part of this secrecy was that many brave men died because the information concerning safety and technical development was not shared.

The accident in October 1960 cost the Russians the experience and leadership of Mitrofan Nedelin. But it also cost the Russian space program a substantial part of its skilled launch vehicle technicians. Due to the political situation surrounding the launch, Field Marshall Nedelin ignored the 20-year-old safety precaution of draining all fuel tanks prior to any repairs on a rocket. The Germans had first set this principle during their rocket tests in World War II.

The physical cause of the failure was the ignition of the fourth stage of the rocket. The human cause of the failure was the decision by Mitrofan Nedelin to perform repairs on a fully fueled rocket. The root cause of this failure is the latent cause, the Russian political imperative to get the launch completed during the visit of Soviet Premier Nikita Khrushchev to the United States.

But this was not the only accident. The Russians lost a cosmonaut to an oxygen-rich fire during on-ground testing in the 1960s. This was a sad event, but there was an even sadder outcome. The news of this oxygen-rich fire was never made public. Had this incident been known, the U.S. space program might not have chosen to use an oxygen-rich environment in the Apollo program. This decision contributed to the loss of the three Apollo 1 astronauts in an on-ground fire in 1967.

Space programs are still not immune to accidents. In August 2003, Brazil lost its one and only launch pad due to an accident. It also lost 21 skilled scientists and technicians, most of that country's knowledgeable space personnel. Here is the Spacetoday.net news report from September 2, 2003:

> Brazilian government officials vowed last week to continue development of a small launch vehicle even after an accident last month claimed the lives of 21 workers. The accident, which took place as technicians were preparing the VLS-1 rocket for a launch attempt, has been seen as a major setback in Brazil's efforts to become a spacefaring nation. Reuters reported that Brazilian President Luiz Inacio Lula da Silva vowed to continue the program and carry out a successful launch by the end of 2006, when his four-year term expires. Repairing the damage to the Alcantara launch facility is estimated to cost $33 million, far more than the Brazilian Space Agency's annual budget of $12 million; it was unclear how much extra funding the government would provide. Brazil's budget is constrained by International Monetary Fund agreements that require the government to run budget surpluses, *Space News* reported.

CHAPTER 5

A Few Pitfalls and More Useful Tools

IMPORTANT ASPECTS of failure investigation addressed in this chapter include:

- Common pitfalls encountered in failure investigations
- Recognizing processes and "quick fixes" that companies often try to substitute for failure analysis
- Understanding important skills and characteristics of professional failure investigators
- Additional resources available for failure investigations

Introduction

The goal of any failure investigation is to discover the failure's root cause(s). The first five steps discussed in Chapter 4 point the failure investigator toward that goal. Once the root cause(s) is determined, the second part of an investigation is to determine the best corrective action(s), as noted in the last four steps discussed in Chapter 4.

As we've seen, three basic tools that are helpful in any failure investigation are a fault tree, a failure mode assessment (FMA) chart, and a technical plan for resolution (TPR) chart. They provide a documented, interchangeable, and concise set of information to ensure that all possible root causes are evaluated logically and efficiently. After determination of root cause, these three tools are then essentially repeated in the form of a corrective action tree, a corrective action assessment (CAA) chart, and a technical plan for evaluation (TPE) chart.

Pitfalls

Pitfalls may be encountered along the way, however, including such common problems and mistakes as:

- Pursuing a "silver bullet" theory
- Not understanding the problem
- Not considering all possible failure causes
- Jumping to conclusions
- Not identifying all the root cause(s)
- Runaway tests and analysis
- Adopting a "remove and replace" company mentality
- Not asking for help
- Returning the part to the supplier
- Failing to follow through
- Not understanding how the failed system is supposed to operate
- Tearing a system apart without a developed plan

The silver bullet theory comes into play when someone involved with the failure investigation is convinced that there is one obvious root cause and the investigation turns into a "prove the theory" exercise. Someone almost always espouses a "pet" root cause. Resist the urge to spend all your time, money, and labor in attempting to prove it. If wrong, you have not only wasted time and money but perhaps destroyed evidence that would lead to uncovering the actual root cause. Not all failure investigations have enough resources and available samples to test every root cause, so you may have to be selective in your approach.

In addition, the failure may be due to a combination of root causes, not just one. Just because you prove the "pet" root cause does not mean there is not another, more important root cause out there. Remember that the goal of a failure investigation is to determine the root cause or causes, not to prove one root cause. When an investigation pursues a singular theory, it has gone off-track. Welds are a great example. Some engineers think that if a failure occurs in a system with a weld, the failure automatically must be due to the weld. They become blind to any other possible root cause.

Not Understanding the Problem. Engineers so often want the details of the problem that they forget to hear the reason their customer has requested help. Information is crucial at this point. Listen to your customers. They may not clearly know what they expect from your investigation, and you may have to help them define your work. The better the definition at the start of the failure investigation, the better the chances the result will satisfy each customer's expectations. In addition, you need to become as knowledgeable as possible about the component, machine, system, or process. If you do not understand what you are trying to investigate, you will never get to the root cause. I have been asked to review failure investigations in progress. While listening to the presentation on the investigation's status, I've asked exactly what it was the customer asked the investigators to do and often have been met with a blank stare.

Not considering all possible failure causes and jumping to conclusions are usually due to the same reason. The failure investigation is moving so fast that the engineers are not taking the time to consider all possibilities. Instead of developing a fault tree, they fall prey to the silver bullet theory. They just tell themselves, "It must be the weld," and move forward to prove it.

Not Identifying All the Root Cause(s). This occurs when brainstorming is not used in the initial stages of the failure investigation or the right people are not present at the brainstorming session. It also occurs when the fault tree is not updated during the failure investigation and has become dormant.

Runaway Tests and Analysis. When investigators become involved with testing and evaluations, they sometimes forget why the tests and evaluations are being performed. The objective is not to run a test and provide the result, but to determine how the result either proves or disproves one of the root causes identified during the initial stages of the failure investigation. In many investigations, the testing becomes an investigation of its own by evaluating the test parameters or variations in the test results. I have read incredible reports that cover the technical aspects of the test performed but forget to discuss how the test results are pertinent to the investigation.

Adopting a "remove and replace" company mentality is common in the fast-paced production world. Why spend time and money to perform a failure investigation when you can just replace the broken part with a new one? The question no one asks is, "What makes you think that the new part is any better than the one that just broke?" It is easier for a program to meet schedule this way, but the company never learns anything about the quality of products they are using. Perhaps it was not the product, but the installation procedure. This kind of mentality is called "Swap till you drop." Eventually you end up with a stockpile of parts you do not want to use. Then you run out, and what do you do?

Not Asking for Help. Companies must demonstrate that it is okay for engineers to ask for help. If this company culture is not developed, the opposite becomes the norm. The result is that engineers perceive asking for help as unacceptable behavior. Everyone needs help eventually.

Returning the part to the supplier is another cost-effective way to deal with hardware failures. This "solution" is popular with company management because the time, labor, and cost of the failure investigation is now the supplier's responsibility. It is very simple to return failed hardware to the supplier and tell them to fix their problem, but what are they going to do? The answer is most often "Swap till you drop, supplier style." Many times it is easier for the supplier to make a new part than to fix the problem. It takes less time and less money, but the root cause is never determined. Returning a part to the supplier is not a failure investigation, it is a pitfall. Nowadays, fewer suppliers have the technical capability, time, or desire to perform any kind of investigation. Once the part is returned, there is a great chance it will simply get lost or thrown away. Worse yet, it may end up back in your next shipment.

Failing to follow through is a pitfall related to the assignment of actions and obtaining agreement on the expected completion date. I cannot stress enough that the person assigned to a task must agree both to perform the test or analysis and to meet the completion date. If someone agrees to the task and the established deadline, then the work normally gets accomplished. As the engineer in charge, it is your responsibility to check on the status of work in progress or create a reporting system to do so. Remind people of their completion date and check on progress. You are running the investigation, so you are responsible for keeping it on schedule.

Not Understanding How the Failed System Is Supposed to Operate. Performing a failure investigation without knowing how the failed system is supposed to operate is like driving a car with three wheels. It may work, but sooner or later you will go off course. Gaining such knowledge takes time, but it is time well spent. For example, not knowing whether a rotating component rotates clockwise or counterclockwise is a simple mistake, but can be very costly when evaluating the crack progression on a fracture surface. Does the automatic welder travel left to right or right to left when making a weld? Does the shaft rotate in the assembly or does the assembly rotate about a fixed shaft? Simply looking at a broken shaft in the laboratory may not provide this kind of information.

Tearing a system apart without a developed plan is a personal favorite. Almost all failure investigations have an established deadline and become a game of "beat the clock." In many cases, the customer has already disassembled and cleaned the failed hardware in order to facilitate the investigation's startup. Unfortunately, this disassembly results in the loss of important information—for example, hardware dimensions, assembly process, tolerance stack-up, fastener torque level, and so on. Cleaning the hardware may destroy even more important information, including debris, corrosion products, and the presence of fluids or lubricants.

Consider, for example, the failure of a small metering pin for an expendable launch vehicle's propulsion system. These pins are very small, about the size of a pencil lead, and are designed to meter propellant flow by moving back and forth into a hole through which the propellant flows. They are made from A286, a heat-resistant iron-base superalloy, and are heat treated to 1100 MPa (160 ksi). A pin had failed, causing the propellant flow to become uncontrolled.

The silver bullet theory for the failure was an overload failure because the material used to make the pin had not been heat treated to the proper strength level. Therefore, in order to determine whether the material and strength were correct, the program manager wanted the pin immediately cut up, placed in a metallographic mount, and microhardness tested.

The engineer in charge of the failure investigation asked for my assistance to persuade the program manager that the best course of action was to first visually inspect the pin and document the inspection. The next step would be to perform a scanning electron microscopy (SEM) examination and documentation of the pin and its fracture surface. The final step would be to cut up the part for metallographic examination because the entire part would have to be sacrificed. The program manager was convinced it was a strength issue due to poor heat treatment and that the pin had failed by overload. I wish I had the crystal ball some program managers seem to have hidden in their desks.

The crazy part of this situation was that it took more time to persuade the program manager than to perform the visual and SEM examinations. These examinations proved the failure was caused by high-cycle fatigue with very little final overload. Evidence of indentation marks on opposing sides of the pin indicated that it had been striking the inner diameter of the hole. The SEM also indicated that the material was correct. All this information would have been lost if the part had been cut up immediately. The metallographic examination and microhardness test were performed and proved the material had been heat treated correctly. The root cause of the failure was determined to be pin design, rather than poor mechanical properties due to improper heat treatment.

Failure Investigation Is Not . . .

Many customers and companies believe there are substitutes for a failure investigation that determines root cause. The belief is that these substitutes can save time and money. In the end, they do neither. Such substitutes include:

- "Give me a five-minute failure analysis"
- "Give me your best guess"
- Reworking or repairing based on no failure investigation
- Scrapping the hardware
- Swapping out for another part, supplier, or something else
- Ignoring the problem and hoping it is a unique occurrence
- Applying band-aid fixes

Five-Minute Analysis. I don't know how many times I've had someone walk into my office and briefly tell me what they think happened and what material they think the part *might* be while holding a bad copy of half an assembly drawing in their hands. Then they inevitably say, "Just give me a five-minute failure analysis." This is an invitation to disaster. You may think you are helping the customer out, but you are putting yourself in peril. Sooner or later, that customer will repeat your answer as fact.

For example, one time I was called to a meeting three months after a twenty-minute telephone conversation with an engineer concerning a failed part. At the conclusion of the call, I asked the engineer to bring me the failed part so I could perform a failure investigation. He agreed, but never arrived. The meeting was a presentation to the customer of the engineer's new design, and I was there to support the presentation. I met the engineer's manager prior to the meeting and told him I was not really aware of the program but I would help in any way I could. The manager looked puzzled. He told me that the design team had redesigned the part based on my failure analysis. I told him I was unaware of any failure analysis. We went outside the room and he showed me the presentation. One chart noted a failure investigation by me, with the conclusions at the bottom of the chart. The manager said he expected me to fill in the details for the customer. I told him there were no details to provide as no failure investigation was ever performed because the design engineer never showed up with the part. The manager canceled the presentation.

There is an old saying, "Believe half of what you read and none of what you hear." If you are at a meeting and someone presents a "fact" about which one of your coworkers supposedly has made a definitive technical statement, be careful. Check it out before providing an opinion on the subject based on the statement. Ask yourself, why was that person not invited to this meeting? I was at a meeting and one of the participants said that he had spoken to a certain coworker of mine the previous month and received an answer to a technical question. Unfortunately, my coworker had left the company more than a year before.

Best Guess. Also be careful when someone says, "Give me your best guess." That is the beginning of a conversation based on a large trust factor. Customers you have worked with before will come to you with a problem, but without the time or budget to get you involved on a substantial level. So they will ask for a best guess. I usually respond by asking at least 20 questions before I provide any kind of technical answer. It drives many customers crazy, but I can usually get enough pertinent information to provide an answer or tell them I do not know. Ideally, the game of 20 questions will impress upon customers the need to do a failure investigation as they come to realize that *their* customers may ask the same questions.

In the course of any given day I answer questions by telephone, by e-mail, in formal meetings, or while walking in the hall. Many times I can answer the questions, but sometimes I say that we need to meet and talk further. This kind of operation is common in job shops or companies that have many small programs.

Companies with proprietary programs have a unique problem. Customers will come to you and say that they cannot tell you anything, but they want your help. When you start to ask a question, they may say, "I can't tell you that." You may not know if you are helping or not, because you do not know exactly what is going on. I recommend that you smile and do the best you can. But always reserve the option to say that there is not enough information provided to determine a proper answer. At times like this, remember the joke about the four blind men and the elephant (Fig. 1).

Four blind men are taken to the zoo. During their visit, the tour guide asks whether they know what an elephant is. The four blind men say no, so the tour guide takes them to meet an elephant. The first blind man touches the elephant's trunk and says the elephant is like a snake. The second blind man touches the elephant's tail and says the elephant is like a horse. The third blind man touches the elephant's leg and says the elephant is like a tree. The fourth blind man touches the elephant's tusk and says the elephant is like a large tooth.

The moral of the story is that if you cannot see the whole picture, then you are going to come to the wrong conclusion and give the wrong answer. In today's fast-paced work environment, you are sometimes forced into that position. When someone tells you to "give me your best guess," be careful, because that person is likely going to take it and run with it. This

Fig. 1 What is an elephant like? Photograph by Daniel P. Dennies, Kenya, 1986

is especially true if you're considered an expert. The customer will make the point that he or she spoke with you, but will not want to pay for a complete investigation.

Rework or Repair without a Failure Investigation. On the production floor, it is sometimes quicker and easier to rework or repair hardware rather than spend time and money performing a failure investigation. Can we just weld it? Can we just put in a piece of sheet metal with a few rivets and submit a waiver to the drawing? Many times, production personnel do not care whether they know the root cause of a failure. The aim is just to skip to the corrective action. If the error is serious enough, then production will request a drawing change to match the rework. For now, fix it, and change the drawing later. The sad part is that the coordination for the drawing change fails. The company orders a whole new set of parts, but the change to the drawing never occurs. The same part shows up and the same failure happens. Or worse yet, the drawing does get changed, the new parts are delivered, and the failure still happens. One of the uses of a failure investigation is to determine root cause and provide definitive corrective action.

Scrapping the hardware is another potential alternative to a failure investigation. It is sometimes difficult to justify a failure investigation of certain parts when it is cheaper to make more. However, how many parts do you have to throw away before it is decided that it's worthwhile to determine the failure's root cause? One per order? Five per order? Ten to twenty parts per 1000? One percent of total yearly sales? The allowable scrap rate is industry and company dependent.

Companies scrap hardware based on various criteria. An obvious scrap is when the hardware fails an established performance criterion, such as mechanical properties or defect level. Another reason may simply be loss in the production system. The criterion may be customer imposed or company imposed. It is only when the scrap rate exceeds the company criteria that the reasons for scrap become a failure investigation unto themselves. Statistics on scrap can require that a failure investigation be performed.

As a company begins to look at the reasons for scrap, many disturbing questions arise. Is the test criterion inadequate? Is the test being performed to evaluate the hardware not appropriate for the material? What is the probability of detection (POD) of the inspection procedure? A POD provides an indication as to how often the inspection procedure will find the defect level specified. The POD indicates how much hardware is being shipped that may not meet the defect level and could fail later. Most companies solve this statistical problem by inspecting for a slightly tighter criterion.

For example, the rocket engine business uses a 5Al-2.5Sn extra-low interstitial (ELI) titanium alloy. At one time, the titanium producers were scrapping 50% of each heat of material due to the difficulty in melting it.

If 50% of each heat is scrapped, just how good is the other half? The company was melting 10,000 lb and shipping 5000 lb. How much confidence do you have in the 5000 lb that is used for your hardware? The titanium producer stated that the product satisfied all specification criteria.

Another example is for an aluminum alloy. As noted earlier, it was discovered that the aluminum producer was pouring six heats to get three heats of 7050 aluminum. Once again, how confident are you in three heats of material that the company receives? But when you start playing the scrap game, and are willing to throw away some percentage, there will be hardware that is obviously bad, hardware that is obviously good, and hardware that might be good or bad due to variations in production and inspection statistics. There always will be a percentage of hardware that may fail during customer use. A good engineering approach is to have adequate inspection on both raw material and finished hardware.

Swapping parts is another poor substitute for a failure investigation. In this scenario, the company discovers a problem with a part and swaps parts from another assembly to keep the assembly line running. When swapping becomes standard practice, tracking all the parts becomes difficult. The company may have stolen so much to get one unit out the door that it ended up using the bad parts from the first unit on another unit. Such parts should be destroyed or marked to ensure bad hardware is tracked. Then the hardware can be repaired for use on another unit. Production lines are always in a hurry. They cannot spend enough time to do it right the first time, but always have enough time to fix it.

What should a company do with parts that are scrapped? At a minimum, the parts should be marked with some type of ID tag and stored in a nonconformance area. From this area, the parts can be controlled to determine which can be salvaged or stripped for pieces. You can steal good pieces off bad parts. It is a way to save hardware, but you have to keep track. Serializing is a good process for keeping track of hardware, but many companies feel it is too time consuming. And the process does not always work. I have performed failure investigations on parts that were supposedly scrapped during production, but somehow made their way into hardware.

Ignoring the problem is an interesting substitute for a failure investigation. It is scrapping without a memory. When a company finds a failed group of hardware, it just scraps the hardware and ignores the problem. The next group will be fine. The company is not going to spend the time and money to do a failure investigation, but instead chooses to see what happens when the next group of parts arrives. By that time, however, everyone has forgotten the high scrap rate on the previous group. Or maybe the next group is fine, but the group after that has a high scrap rate.

When a failure investigation is finally performed, the engineer must track the history of these events because there may not be any documentation of previous failures. Ask these questions: "How long has this been going on?" "Is this the first time the failure has happened, or has this happened before?"

Band-aid fixes are another quick form of corrective action based on no failure investigation. The band-aid does not solve the problem, but makes the part usable for its application.

Consider, for example, hot isostatic processing (HIP) of titanium castings. The process has been shown to close internal casting flaws. But a flaw cannot be open to air, because the oxide formed on the surface will not allow the flaw to close and diffusion bond during the HIP process. The HIP process will close the open flaw, but it will not diffusion bond together with the surface oxide. The result is a knife-line defect that cannot be detected during inspection. As a band-aid fix to recover hardware, some suppliers weld over the cracks open to the surface and then HIP the casting. The parts pass inspection, but fail in service.

Another example involves a stainless steel casting. Because of their very poor fluidity, stainless steels do not cast well. One day, the production area was machining a part fabricated from a 17-4PH martensitic stainless steel casting. During machining, a straight-line crack was observed. Straight-line cracks are not normal in castings. As the machining of the part progressed, two more straight cracks 90° to the first crack and parallel to each other appeared. It turned out that the casting was so poor that the casting shop had machined out a bad section and welded in a plate. When making our part, we machined through the weld and into the separation of the casting and the plate. The interesting conclusion to the story is that we ended up using the part. Because we needed the part so badly, we welded it again to produce a band-aid fix of our own.

Unfortunately, this example does not describe a new concept. During the construction of tall buildings in the 1920s and 1930s, welders were caught putting bar stock into large multipass welds in the main structural supports. Welders were paid by the linear feet of weld they made each day. When faced with a 30-pass weld, it was much quicker to throw a bar into the middle and weld over the top to fill the gap. Failure investigations of collapsed buildings would discover the welds with the bar stock in the middle.

Summary: Remember, in the end, these substitutes are generally neither cost effective nor schedule friendly.

What Can You Do?

So far this book has shown you how to:

- Comprehend the complexities of organizing a failure investigation and understand the importance of initially defining a clear and concise goal, direction, and plan

- Recognize the many aspects and organizational levels that may contribute to and define a failure
- Understand the requirements of leading a successful failure investigation
- Become an advocate for the discovery of a failure's root cause(s) through the use of a well-organized investigation
- Understand the methodology for an organized failure investigation
- Understand some of the pitfalls if you don't remain organized

In addition, a professional engineer who performs failure investigations must always strive to improve these skills:

- Technical skills
- Communication skills
- Technical integrity

Technical Skills

You will be become more technically proficient with every failure investigation because each will require you to expand your technical knowledge. This book is a good start for improving your technical skills. Such skills can also be sharpened and honed through courses provided by technical societies and colleges, books, conferences tailored for specific topics, mentoring from within your company or organization, and the Internet.

The Internet has become the sole source of information for many young engineers. It is accessible and quick, but please be aware that not all technical information on the Internet has been validated or screened, and it is not difficult to find instances where the information is incorrect. In addition, most information developed prior to 1990 is not posted on the Internet—so many good technical papers written before then will not be found. Many younger engineers go straight to the Internet because it is the source with which they are most familiar. On the other hand, many older engineers do not use the Internet at all. Use all available resources.

Communication Skills

Communication skills can be separated into report-writing skills, presentation skills, and computer skills. Many engineers lack proficiency when it comes to writing a report or making a presentation and are intimidated by the prospect of doing either. Younger engineers usually have proficient computer skills but lack presentation skills. The situation is reversed for many older engineers.

Writing Skills. The ability to write a clear, concise report is a learned talent. Most colleges and universities have dropped writing courses for engineers and hope they learn this skill in their technical courses.

Initial Memo. In every failure investigation, start by writing a one-page memo to establish four items negotiated at the onset: the investigation's priority, the resources available, any constraints imposed, and the goal. This memo serves to document agreement on the negotiated items. All parties involved immediately start the investigation with the same knowledge and expectations. If there is a misunderstanding concerning one or more items, it can be corrected immediately, not weeks later. Make this a habit for all failure investigations. As an option, some companies use a letter of agreement.

Status Reports. As the failure investigation progresses, send out status reports. Establish with each customer what the expectations will be regarding reporting, presentations, e-mail contact, telephone contact, and so on. Establish the time frame of the status reports and what form the reports will take. Even if only using e-mail, send the fault tree, FMA, and TPR charts discussed in Chapter 4. Status reports and e-mails should be short and concise, no longer than one page, and always dated. Keep them simple; explain the changes from the last status report and welcome questions. Follow up as necessary. The last thing you want to hear from your customer is, "Where have you been, what are you doing, and why are you keeping us out of the loop?" The status report is a very simple way to keep them informed. If they want something more formal, discuss format based on requirements.

E-mail is an easy way to send quick responses. Sometimes it is too quick and too easy. Many people have sent e-mails they wish they had never sent. Do not respond to negative e-mail immediately. Wait a day, even if you receive flame-mail. (You probably have seen this kind of e-mail. It's usually typed in a bold, red 32-point font.)

It is easy to get caught up in e-mail discussions, but after a while they lose their effectiveness. At some point you have to decide to switch from e-mail to the telephone to an in-person conversation. The best approach is to just e-mail the statement, "This is important, and we need to discuss it in a meeting." Do not get lured into being an e-mail "answering machine." Lastly, be careful when responding to questions via e-mail, because there may be bigger issues behind simple questions.

Final Report. Once the failure investigation has been completed, always write a final report. If you wait too long and let the report get cold, the clarity and importance of the investigation will become lost. Do not let this happen. Not only does your customer expect a final report, but it's necessary for the statistical database you are creating.

When a customer does not like the outcome of an investigation, they may not want a final report. That position is not good for them or for their company. Negotiate in the beginning that there will be a final report. Stipulate the required hours to complete the final report in the investigation's budget. If the customer attempts to stop paying, you have documentation that they agreed to a final report.

Summarize pertinent information on the first page of the final report. Think of the report as a newspaper article, where the first paragraph is important. Communicate clearly and concisely. After writing the report, give it to two people for review: another expert in your field who can review the technical content, and a nontechnical person, preferably someone with good writing skills, who can make sure the report reads well to a less technical person. Anyone should be able to read the report and understand the final conclusion.

Presentation Skills. There is only one way to improve your presentation skills: Perform presentations. Very few natural orators enter the engineering profession. Everyone is nervous, we have all been through it, but you have to make presentations in order to hone your skills.

Some engineers were introduced to presentations at university. Graduate students usually get more practice, particularly at conferences. Practice at home in front of a mirror. Look forward and maintain eye contact. Videotape your presentation and play it back at a very fast rate; that way you will notice distracting hand or head movements. You can also learn by watching other people and the tools they use during presentations. To emphasize a point, walk to the screen and put your hand on it, or walk toward someone who has asked you a question. Practice your presentation in front of a "friendly" audience, such as your coworkers, before actually presenting it to the customer. Once you gain a bit of confidence, it becomes easy and even fun.

After becoming comfortable with presentations, move from the familiar territory of your company to the not-so-familiar territory of conferences. This will provide good feedback and new ideas. You'll learn that a conference presentation can also serve as a "fact-finding mission." By presenting a technical issue and the work you have done, you may receive help and feedback from people who have had similar problems.

Computer Skills. Computer technology is rapidly replacing paper, as drawings become datasets and interoffice memos become e-mail. Unfortunately, computer hardware and software keep changing. You must stay current. The best person to turn to for assistance is the latest new hire out of college. Most companies offer computer classes, as do local schools. If you have children in high school or college, they may also be very good resources.

Technical Integrity

Engineers in the workplace environment have more pressure today. Companies continually demand more risk acceptance by first-level engineers and, in many cases, jobs may be on the line. Companies are looking for ways to get things done quickly and they need technical signatures. Therefore, pressure is put on engineers. I would propose that you remember a saying I heard once, "The right thing and the expedient thing are rarely the same thing."

Consider another basic fact. Engineers are beginning to be sued on an individual basis. Previously it was companies and corporations that were sued when products failed. Now lawyers are suing the presidents of the companies and the engineers that signed drawings. Another interesting fact is that if engineers are involved with a lawsuit based on their work, companies may not offer legal support.

If you get in a situation where you have to defend what you did and you can demonstrate good organization, good procedure, good rationale, and good documentation, you have the information necessary for a strong case. Lawyers do not usually want to talk to someone who is prepared and organized.

The following discussion on integrity is adapted from an item posted on a church website (Ref 1).

The Test of Integrity

Have you ever heard of a "secret shopper"? Contrary to what you might think, a secret shopper is not a person who goes on wild buying sprees and then hides the receipts and credit card bills. A secret shopper is an individual employed by [a retail company] to test the integrity of its salespeople.

I have a friend whose business does millions of dollars a year in retail sales, both locally and nationally. He's decided to hire some secret shoppers. When you hear his story, you'll understand why.

In the 20 years he's been in business, my friend estimates over $500,000 lost to employee theft—maybe as much as $1 million. Last winter he discovered that an employee in another state had ripped him off to the tune of $80,000 in cash. After firing her, he took the woman to court in an attempt to get the money back. As you might expect, it's all gone. The judge ordered her to pay it back at a rate of $50 a month, which, if you do the math, means my buddy will get his money back in about 133 years.

So, he's hiring secret shoppers. Their job is to go into a store, make a purchase when no other customers are around, and pay the exact amount due in cash. But the trick is, they don't wait for the employee to ring up the sale. They just put the cash on the counter, say "I'm in a hurry," and leave the worker alone with the money.

Some people ring up the sale as usual and put the money in the register. Others pocket the cash. A check of the register tapes at the end of the day reveals their actions, and the ones without integrity are history—they lose their job.

Make no mistake about it, friends. Though it may seem that an old-fashioned value like integrity just isn't in style anymore, nothing could be further from the truth. Integrity is vitally important. It's indispensable. Every person who hopes to experience lasting success in any area of life—in a job, a friendship, a business deal, a marriage—must pass the test of integrity.

Of course, it would help to know exactly what we're talking about when we use that word. So, I looked it up in several dictionaries. One defined integrity as honesty, sincerity, and uprightness. Another said it was adherence to a code or standard of values. Still another defined it as moral or ethical strength.

While those are excellent definitions, the more I thought about it, the more I became convinced that the best way to define integrity is in terms of a question.

And that question, very simply, is this: "Can I be trusted when no one is looking?"

Can I be trusted to not steal from my employer when I'm alone in the store? Can I be trusted to honor my marriage vows when I'm by myself on a business trip? Can I hear the deep dark secret of a friend and be trusted not to blab it to the world behind their back? Can I be trusted to keep my eyes on my own test paper when the teacher leaves the room? Can I be trusted to turn in expense reports that aren't padded with make-believe expenditures? Can I be trusted to be where I've told my parents I would be? Can I be trusted to surf past the sexual immorality on the Internet or cable when no one else is around?

Can I be trusted when no one is looking?

That question reveals the degree of integrity that is present in our character. And our answer to it is a reliable predictor of our future success or failure.

Notice that the issue is not "do others trust me?" That's a measure of image, not integrity. The issue is, when I examine what's behind the veneer of the image that I project when others are watching, do I really deserve their trust? That's the true measure of integrity.

Most engineers enjoy a long career and build strong, ethical, honest reputations. But if an engineer gets caught in a lie just once, his or her career is changed forever. It may even be over. When engineers are labeled as being unreliable, no one will believe them and they are no longer sought out for information.

Politicians provide a good example. Some politicians believe they should tell us what they think we want to hear, even if it is a lie. They spin the story to make it sound better, even if it is a lie. Then they are shocked when the public discovers the lies, is outraged, and does not support them. These politicians do not seem to understand that the public is more outraged by lies than by any information that is the truth.

Other Useful Tools

This book is a tool for organizing a failure investigation. I've developed the approach presented here and find it to be a cost-effective technique that works extremely well in most situations. The same approach can be applied to most problems. However, a toolbox contains many tools, and each of us needs to find the right tool for the way we operate, or the right tool for the right job. Every failure investigation is unique. Therefore, it pays to have many tools in your arsenal.

Other tools for failure investigation and root cause determination include processes and methods developed by Kepner-Tregoe, Inc. (Ref 2), The Reliability Center, Inc. (Ref 3), The Failsafe Network, Inc. (Ref 4), and Shainin LLC (Ref 5).

The Kepner-Tregoe (KT) method streamlines the problem-solving technique used in failure analysis and works well in industrial failure situations. The KT method is not as thorough as fault tree analysis, because potential causes get trimmed early in the process if they are found to be impossible. However, it is a useful method of problem solving and decision analysis.

KT is a structured problem-solving approach that first defines the problem in terms of what "is" and what "is not" as related to the questions of what, where, when, and extent. For example, gas turbine blade cracking may have been confined to one type of cracking mode, one specific row of blades, and one type of turbine blade design, with the problem occurring only after a certain amount of operating hours and observed in all plants. This shapes what "is" and what "is not" in a KT session. Once the problem has been defined, possible root causes are developed and compared with the "is" and "is not" problem statements to test if they are possible. Once impossible root causes have been eliminated, the possible root causes are ranked in terms of probability. In real world practice, additional investigation focusing on the remaining possible causes is then performed to try to eliminate all but the one true root cause.

The Reliability Center has developed a software package called PROACT, a comprehensive tool that automates the root cause analysis (RCA) investigation process. PROACT software allows you to concentrate on solving the problem at hand instead of having to worry about where you are going to keep all your data and how you are going to present it. PROACT serves as a project management tool.

The software is purported to provide a number of user benefits. PROACT incorporates RCA methods that have been proven to reduce production costs while increasing production output. It cuts analysis time in half by eliminating the extensive administrative work that is usually associated with conducting RCA. It provides a standardized reporting format so that users and decision makers can focus on the report's content and not its format. It provides a presentation module that allows the analyst to present critical information to decision makers quickly, clearly, and easily. It maintains all analysis data so that you do not have to search through volumes of computer and paper files to get the information that you need. Lastly, it provides an easy-to-use drag-and-drop "logic tree" that focuses on solving the problem and not on becoming a CAD operator.

The Failsafe Network has promoted various concepts in failure investigation. Operation Failsafe is a comprehensive plan for ingraining the root cause mentality within people and their organizations. The ROOTS Investigative Process is a flexible, open-ended, practical approach that helps discover the physical, human, then latent causes of failure. An evidence-driven process, it proposes to drive the serious inquirer into areas most in need of clarification. The WHY Tree is the "inquiry engine." It proposes to be a tool for keeping investigative teams on track by documenting where you are in your current understanding of a problem, whether it be with a team or as an individual.

The Combing Process is a failure mode and effects analysis (FMEA)-based tool for rapidly identifying your "significant few" chronic failures in hours or days. It proposes to be an approach that goes to the hands-on operating and maintenance people for information, rather than relying on existing databases. The Situation-Filter-Outcome Model proposes to be a practical and decisive way to put yourself in someone else's shoes—to understand why people do what they do. It can help to bring to the surface the latent causes of a failure. The Significant 6 Barriers to True Root Cause Discovery is a series of statements, articles, and graphics intended to make the real issues visible. Lastly, the Root Cause Conference is a nonproprietary, intercompany, interindustry body of people acting as a beacon for exposing, then acting on, the real root causes of things that go wrong in business and industry.

Shainin is a consulting and training firm whose sole purpose is to help manufacturing and product development businesses avoid and reduce costs associated with technical problems. They partner with clients to help them apply Shainin management, engineering, and manufacturing strategies to effectively reduce costs, add value, and improve operational efficiency. One key to their success is a constantly growing set of proprietary strategies that they have spent 50 years creating.

Shainin uses a statistical engineering approach to technical problem solving. They propose three distinct paths to help train personnel in precise areas of expertise:

- ROLLING TOP 5 for improved business performance
- GREEN Y for problem prevention
- RED X for problem solving

Summary. Find the tools that work for you. Take the methodology presented in this book and alter it to suit your needs. Once you start to use the tools offered, you will have an organized failure investigation that will impress both your customers and your management. The more you work with these tools, the faster and easier the process will become.

REFERENCES

1. "The Test of Integrity," North Heartland Community Church, Kansas City, MO, May 11, 1997, www.northheartland.org/1997/051197m.htm (accessed Feb 2005)
2. Kepner-Tregoe, Inc., Princeton, NJ, www.kepner-tregoe.com (accessed Feb 2005)
3. The Reliability Center, Inc., Hopewell, VA, www.reliability.com (accessed Feb 2005)
4. The Failsafe Network, Inc., Montebello, VA, www.failsafe-network.com (accessed Feb 2005)
5. Shainin LLC, Livonia, MI, www.shainin.com (accessed Feb 2005)

Case History: *Andrea Doria-Stockholm* Collision Off Massachusetts (1956)

Sally Van Duyne

The Andrea Doria *sank after colliding with another liner, the* Stockholm, *off Massachusetts.*

Background

On July 25, 1956, the liners *Andrea Doria* and *Stockholm* collided near the Nantucket Lightship off Massachusetts. The *Andrea Doria* sank, with the loss of 43 passengers and crew, the first big liner to be lost in peacetime since the *Titanic* went down in 1912. On the *Stockholm* three crew members were never seen again, and several later died from their injuries. The collision was caused primarily by misinterpretation of radar signals, and the sinking was a result of faulty ballasting.

The *Andrea Doria* was the flagship of the Italian Line. Built by Ansaldo of Sestri, near Genoa, the vessel was launched in 1951, the first passenger ship to come off the ways in Italy after World War II. Genoa was its home port. The *Doria* was 213 m (700 ft) in length, weighed 29,083 tons, had a service speed of 23 knots, and was known for its graceful lines. On what was to be its last voyage, the ship left Genoa on July 17, 1956, and was due in New York City on July 26.

The *Stockholm,* owned by the Swedish-American Line, was built in Sweden by Götaverken of Göthenberg and launched in 1943. It was 175 m (575 ft) overall and 12,644 tons and had a service speed of 19 knots. On July 25, 1956, the ship was on its first day out, beginning its homeward voyage from New York City to Sweden.

Both ships were heading toward the Nantucket Lightship, a vessel anchored 50 miles off Nantucket, to guard the shoals off that island. The Nantucket Lightship is an important focal point for the sea lanes of the world. Although the shortest great circle—a circle outlined on the Earth's surface, such as the Equator, by a plane passing through the Earth's center—track for ships traveling between New York and European ports would be to the north of the vessel, they must pass to the south of it before turning north. Certain shipping lines had agreed to having their eastbound ships pass in a track 20 miles south of the light vessel and their westbound ships pass close to it. However, neither the Italian Line nor the Swedish-American Line was a party to this agreement.

Details of the Collision

On the afternoon of July 25, 1956, the *Andrea Doria,* eight days out of Genoa and due to dock in New York the next morning, encountered fog when it was still about 241 km (150 miles) from the Nantucket Lightship. Captain Piero Calamai observed the normal precautions for these conditions: closing the watertight doors, sounding regular prolonged blasts on the siren, having an officer keep a constant watch on the radar screen for any sign of approaching ships, and reducing the ship's speed. Only the last of these can be criticized, as the *Doria* had reduced its speed only 5%, from 22 knots to 21 knots, and the "moderate speed" called for in the collision rules is defined as a speed that allows the ship to be stopped within its visibility distance. By 8:00 A.M., the *Doria*'s visibility was only 8 km (0.5 mile), and stopping her would take considerably more distance than that. However, it was the custom among sea captains bound by schedules to make only token reductions in speed in bad weather, while keeping a close watch on their radar.

At 10:20 A.M., *Doria* passed 16 km (1 mile) to the south of the light vessel. At about 10:45, Second Officer Franchini saw a "blip" on the radar screen that represented a ship that appeared to be not quite due ahead, but fine, or slightly, to starboard. Captain Calamai checked the radar screen himself and judged that the rapidly approaching ship was to starboard—that is, to the right—of his heading marker. Captain Calamai felt quite certain that the two ships were on a starboard-to-starboard path, but thought that the distance between would be uncomfortably small. He gave the order to alter course to four degrees to port (left). In so doing, he was going against rule 18 of the collision rules, which states that "when two power-driven vessels are meeting end on, or nearly end on, so as to involve risk of collision, each shall alter her course to starboard, so that each may pass on the port side of the other." He had two justifications for going against this well-known rule: he believed that the two ships were clearly on a starboard-to-starboard course, and he thought that an alteration of course to starboard might force his ship into the shoal water to the north of the light vessel.

When the two ships were less than 8 km (5 miles) apart, the *Doria* still could not see the *Stockholm*; the *Stockholm* could see the masthead lights, but not the sidelights denoting port and starboard, on the *Doria*. When the ships were about 3.2 km (2 miles) apart—and at their combined speeds of 40 knots would close in on each other in three minutes—the *Doria* could discern the first navigation lights on the *Stockholm,* and could see that they were not green as expected in a starboard-to-starboard passing, but red. The red lights meant that the *Stockholm* was crossing the *Doria*'s bow to make a port-to-port passing. At that moment Captain Calamai gave the fatal order, "Hard-a-port," which was consistent with his plan for a starboard-to-starboard crossing. There was nothing the crew could do as they watched the bow of the *Stockholm* plow into their starboard side.

The weather encountered by the *Stockholm* as it traveled east toward the Nantucket Lightship was very different from that encountered by the *Andrea Doria*. Somewhat before 11:00 P.M., as the *Stockholm* traveled through a clear moonlit and starlit night, Third Officer Carstens-Johannsen first saw the sign of an approaching ship at the 19 km (12 mile) radius on the radar screen. He recorded the position of the approaching ship as two degrees to port of his heading marker. Another plot of the ship a short time later also showed the ship to be slightly to port. As there was no sign of fog, Carstens, as he was called, saw no reason to call Captain Nordensen up to the bridge. But he did post a lookout on the port wing of the bridge to spot the approaching vessel, at which time he planned to alter his course to starboard to make a safe port-to-port passing. When the bridge lookout suddenly called out "Lights to port," Carstens immediately ordered a change in course of 20 degrees to starboard. Then he went out onto the bridge and saw to his horror that an immense liner was about to cross his bow, presenting the green lights of her starboard side. He shouted "Hard-a-starboard!" to the helmsman. As Captian Calamai had tried at the last moment to increase the distance in a starboard-to-starboard passing, Third Officer Carstens-Johannsen tried to increase the distance in a port-to-port passing. Then Carstens pulled the telegraphs to "Stop" and "Full Astern."

The stem of the *Stockholm,* stiffened for travel through the ice in northern oceans, plowed into the side of the *Andrea Doria* just below the bridge down to her starboard fuel tanks with a momentum of 30 million foot-pounds. Yet only one main compartment of the *Andrea Doria* was damaged, and it would have been able to withstand the collision had it not been for the severe list that developed soon afterward. Since the *Doria* was near the end of its journey, the starboard fuel tanks contained only air. When the starboard tanks were rent in the collision, water rushed to fill them, making the starboard side much heavier, and causing a list to starboard of 18 degrees and soon after, 22 degrees. A narrow tunnelway that led to a pump room containing valves for flooding the tanks with sea water for stability ran between the fuel tanks on the port and starboard sides. Since this tunnel was so far inboard, a watertight door had not been fitted to it. The tunnel flooded quickly after the collision, and it was no longer possible to get to the valves to flood the port tanks to make them as heavy as the starboard tanks.

When the port side lifeboats were readied, it was found that they were fitted with davits—the cranes that project over sides of a ship—that could work only if the vessel's list was less than 15 degrees. That meant that the port lifeboats were useless; the starboard-side lifeboats could not accommodate everyone and were hard to climb into because, with the list, they swung far out over the water (Fig. 2).

A number of nearby ships rushed to the rescue, working against time as the *Doria*'s list gradually increased and it was more and more likely to capsize. Of the 1706 passengers and crew on the *Doria*, 1663 survivors were taken off; most of the dead were killed on impact. The surviving passengers were all off by 4:00 A.M. on June 26; at 5:30, when the list reached 40 degrees, the officers also left. At 10:09 A.M., 11 hours after the collision, the *Andrea Doria* rolled over onto its side.

Fig. 2 The *Andrea Doria* listing heavily before sinking. The ship's severe list (18 degrees, then 22 degrees) rendered the port-side lifeboats useless and caused the starboard-side lifeboats to swing far out over the water, making them hard to access

Impact

In the legal proceedings that followed, each line claimed that the other line was totally to blame and that its own officers were entirely blameless. The Swedish-American Line sued the Italian Line for four million dollars and the Italian Line sued the Swedish-American Line for 1.8 million dollars. The Swedish-American Line's case was by far the stronger. Most important was the immediate 18 degree list of the *Andrea Doria,* which put the responsibility for the capsizing and sinking onto the owners of the *Doria,* as did the *Doria*'s turn to port to make a starboard-to-starboard crossing, thus violating the Rules of the Road. Also at issue were the speed of the *Andrea Doria* in fog, the failure of its crew to plot the bearings of the approaching *Stockholm,* and the disappearance of the *Doria*'s ship's log.

The elements of the Italian Line's case were Third Officer Carstens-Johannsen's delay before making a starboard turn, his youth and inexperience, and the fact that the *Stockholm* was far to the north of the recommended track for eastbound vessels.

After the hearings had gone on for about three and a half months, the two lines came to an out-of-court settlement, dropping their claims against one another. And so the court never did apportion blame for the collision. It seems likely, however, that the two ships were not on safe parallel courses in which they would pass either port-to-port or starboard-to-starboard, but were converging at a very small angle. A lesson relearned in the disaster is that in order to conform to the collision rules, officers of the watch should treat fine bearings as head-on collisions, and should make large alterations clearly discernable to the approaching ship in ample time.

There have been considerable changes in both officers' understanding of radar and in the technology of radar since the *Doria-Stockholm* collision. Captain Calamai probably did not understand that a four degree change in course would not make his intent to make a starboard-to-starboard passing clear to the *Stockholm*. However, today, largely as a result of this disaster, officers on ships are required to take courses in radar. In addition, by international convention, automatic radar plotting aids have had to be fitted in all large ships since the 1980s. Apparently, no plot was made on the *Andrea Doria* of the *Stockholm*'s bearings, and although one was believed to have been made on the *Stockholm* of the *Doria*'s bearings, it could not be produced, presumably because it had been rubbed out.

The true motion radars in use today allow the officer of the watch to see the actual course and speed of another ship, instead of the bearing relative to a heading marker. The latter system could result in errors because a ship can get a few degrees off course at any time, which may explain why the *Andrea Doria* thought the *Stockholm* was fine to starboard and the *Stockholm* thought the *Andrea Doria* was fine to port.

Another lesson learned from the collision involved the ballast tanks. The designers of the *Andrea Doria* had followed the practice common at the time of obtaining sufficient stability to satisfy the seven degree limit on listing by stipulating that the oil tanks when empty be flooded with sea water. This was an unreasonable burden to lay on the engineers, who did not want to contaminate their fuel tanks with sea water. For in order to clean the tanks, they then would have to pump the contaminated oil and water onto barges for removal ashore. The lessons learned here were the advisability of having separate tanks for ballast water needed for stability, and of having the capability of remote control of any valves needed for counterflooding of the ballast tanks. The 1960 Safety of Life at Sea (SOLAS) convention criticized the practice of using oil tanks for ballast, and recommended that full stability information be issued to the staff and that diagrams of ballasting arrangements be displayed on board.

A final effect of the *Andrea Doria-Stockholm* collision was the introduction of a traffic separation scheme for ships passing near the Nantucket Lightship, which was introduced in 1977. It required that eastbound ships keep to a lane south of the light vessel, and that westbound ships keep to a lane north of the light vessel. There is a separation zone 4.8 km (3 miles) wide, with the light vessel in the middle, between the two lanes.

SELECTED REFERENCES

- K.C. Barnaby, *Some Ship Disasters and Their Causes,* A.S. Barnes, 1970
- W. Hoffer, *Saved: The Story of the Andrea Doria—The Greatest Sea Rescue in History,* Summit Books, 1979
- J. Marriott, *Disaster at Sea,* Ian Allan, 1987
- A. Moscow, *Collision Course,* Grosset & Dunlap, 1981

Case History Discussion

Forty-four years after the sinking of the *Titanic,* on July 25, 1956, two ocean liners, the *Stockholm* and the *Andrea Doria,* collided in the fog near the Nantucket Lightship off Massachusetts. The loss of life, with only 43 dead out of 1706 passengers, was not as great as in the *Titanic* disaster, but the *Andrea Doria* was lost in 400 km (250 ft) of water 11 hours after the collision. As in 1912, shipping lanes, ship speed and design, and the decisions of the captain played a vital role in the collision. A new factor was the use of radar.

The root cause of this failure appears to be human. The *Andrea Doria* bridge crew was criticized for violating the "rules of the road" in trying to make a starboard-to-starboard crossing, proceeding at too high a speed in the fog, failing to plot the approaching *Stockholm* radar bearings, and losing the ship's log. The *Stockholm* bridge crew was criticized for their delay in making a starboard turn, having a young and inexperienced officer in charge, and traveling far north of the recommended eastbound shipping lanes.

The ramifications of this failure were that ship officers were required to take courses in radar and large ships were required to have automatic plotting equipment. The shipping lanes were changed so that all westbound ships were north of the lighthouse while all eastbound ships were south of the lighthouse. The collision also resulted in changes to the design of ship ballast tanks. Separate tanks for ballast water and remote-controlled valves for counterflooding of the ballast tanks became more prevalent in ship designs. The interesting point is that these changes did not occur until the late 1970s and early 1980s, more than 20 years after the collision.

APPENDIX 1

General Procedures for Failure Analysis*

Revised by Dennis McGarry

The objective of all manufacturing is to produce an economical, reliable product whether it is an automobile, an airplane, a tractor, or a toaster. Product quality is inescapably linked with user satisfaction, and today's buyer is demanding more reliability for the dollar than ever before. In order to maintain a competitive advantage through satisfied customers, producers must avoid unreasonable failures of their products in service. In addition, we have become a highly litigious society in which product liability claims are aggressively pursued in the courts as a means to force producers to make safety and reliability improvements and to punish companies financially for unsafe or substandard products. Therefore, analysis of failures that are or could be destined for court have an added constraint of following Federal Rules of Evidence during investigation of a failure.

While it takes engineering teamwork to create a safe product with a superior advantage over a competitor's product, a vital ingredient in attaining this goal is a scientifically based method for analyzing the inevitable failures that occur during engineering tests or in service. Results of field-failure analyses must be fed back to the appropriate design and/or manufacturing personnel to form a closed loop, providing for continual improvement in product reliability.

This appendix is concerned primarily with general procedures, techniques, and precautions employed in the investigation and analysis of metallurgical failures that occur in service. The stages of investigation are discussed, the various features of the more common causes of failure

---

*Reprinted from *Principles of Failure Analysis* (Education Course), Lesson 1, ASM International, 2002

characteristics are described, and several of the fundamental mechanisms of failure are explained.

Information on procedures and techniques specific to the analysis of failures by various mechanisms and related environmental factors, failures of principal product forms, and failures of manufactured components and assemblies are covered in subsequent lessons.

The reasons for component failures are usually related to problems of service conditions, design, material, and its specification, processing, and assembly, and they are often interrelated.

Stages of a Failure Analysis

Although the sequence is subject to variation depending on the nature of the failure and the availability of physical evidence or background information, there are stages that are common to all successful failure analyses. The combination of these stages comprises the total investigation and analysis. The following list includes many of the commonly used stages. The sequence in which these stages are used is not critical. Not all of the stages will or can be used in every failure analysis. The remainder of this lesson describes these stages in detail.

1. Collection of background data and selection of samples
2. Preliminary examination of the failed part (visual examination and record keeping)
3. Nondestructive testing (NDT)
4. Mechanical testing (including hardness and toughness testing)
5. Selection, identification, preservation, and/or cleaning of critical specimens
6. Macroscopic examination and analysis fracture surfaces, secondary cracks, and other surface phenomena)
7. Microscopic examination and analysis of fracture surfaces
8. Selection, preparation, examination, and analysis of metallographic sections
9. Determination of the actual stress state of the failed component
10. Determination of the failure mode
11. Chemical analyses (bulk, local, surface corrosion products, and deposits or coatings)
12. Application of fracture mechanics when warranted
13. Testing under simulated-service conditions
14. Consulting with experts in other disciplines
15. Synthesis of all the evidence, formulation of conclusions, and writing a report (including recommendations)
16. Follow-up recommendations

Time employed in ascertaining all the circumstances of a failure is extremely important. When a broken component is received for examination, the investigator is sometimes inclined to prepare specimens immediately without devising an investigative procedure. To proceed without forethought may destroy important evidence and waste time.

In the investigation of failures, it is always preferable for the analyst to visit the scene, but, for the analysis of some components, it may be impractical or impossible for the failure analyst to visit the failure site. Under these circumstances, data and samples may be collected at the site by field engineers or by other personnel under the direction of the failure analyst. A field failure report sheet or checklist can be used to ensure that all pertinent information regarding the failure is recorded.

In cases that involve personal injury or will most likely involve legal pursuit of compensation from another company, care must be taken in preserving the scene and physical evidence. Accidental or deliberate destruction of evidence can result in diverting the legal liability of a failure to the person or company destroying the evidence, even though they may not have caused the original failure.

The goal of every failure analyst is to determine not only the fracture mechanism that results in the failure but also the root cause of the failure. Often the root cause is related to misuse, poor maintenance practices, improper application, or use of the product in a manner not intended by the manufacturer. The final result may be a material failure, but the root cause may not be related to the material properties, design, or manufacture of the product.

1. Collection of Background Data and Samples

The failure investigation should include gaining an acquaintance with all pertinent details relating to the failure, collecting the available information regarding the design, manufacture, processing, and service histories of the failed component or structure, and reconstructing, insofar as possible, the sequence of events leading to the failure. Collection of background data on the manufacturing and fabricating history of a component should begin with obtaining specifications and drawings and should encompass all the design aspects of the failed part as well as all manufacturing and fabrication details—machining, welding, heat treating, coating, quality-control records, and so forth. Also, obtain pertinent purchase specifications—Are they appropriate, does the material meet the specs? And are the service specs appropriate for the service conditions?

Service History. The availability of a complete service history depends on how detailed and thorough the record keeping was prior to the failure. A complete service record greatly simplifies the assignment of the failure analyst. In collecting service histories, special attention should be given to environmental details, such as normal and abnormal loading, accidental overloads, cyclic loads, variation in temperature, temperature gradients, and operation in a corrosive environment. In most instances, however, complete service records are not available, forcing the analyst to work from fragmentary service information. When service data are sparse, the analyst must, to the best of his or her ability, deduce the service conditions. Much depends on the analyst's skill and judgment, because a misleading deduction can be more harmful than the absence of information.

Photographic Records. Photographs of the failed component or structure are often critical to an accurate analysis. A detail that appears almost inconsequential in a preliminary investigation may later be found to have serious consequences; thus, a complete, detailed photographic record of the scene and failed component can be essential.

Photographs should be of professional quality, but this is not always possible. For the analyst who does his own photography, a single-lens reflex 35 mm or larger camera with a macro lens, extension bellows, and battery-flash unit is capable of producing excellent results. It may be desirable to supplement the 35 mm equipment with an instant camera and close-up lenses.

When accurate color rendition is required, the subject should be photographed with a color chart, which should be sent to the photographic studio for use as a guide in developing and printing.

Some indication of size—such as a scale, coin, hand, and so forth—should be included in the photograph.

Samples should be selected judiciously before starting the examination, especially if the investigation is to be lengthy or involved. As with photographs, the analyst is responsible for ensuring that the samples will be suitable for the intended purpose and that they adequately represent the characteristics of the failure. It is advisable to look for additional evidence of damage beyond that which is immediately apparent. For failures involving large structures or key machinery, there is often a financially urgent need to remove the damaged structure or repair the machine to return to production. This is a valid reason to move evidence, but a reasonable attempt must be made to allow other parties who may become involved in a potential legal case to inspect the site. All concerned parties then can agree on the critical samples and the best way to remove them. If all parties are not available, care must be taken not to damage or alter critical elements to avoid spoliation of evidence. Guidelines governing sample collection are covered in ASTM specifications, E 620, E 678, and especially E 860 and E 1020. It is also recommended that samples be taken from other parts of the failed equipment as they may display supportive damage.

It is often necessary to compare failed components with similar components that did not fail to determine whether the failure was brought about by service conditions or was the result of an error in manufacture. For example, if a boiler tube fails and overheating is suspected to be the cause, and if investigation reveals a spheroidized structure in the boiler tube at the failure site (which may be indicative of overheating in service), then comparison with an unexposed tube will determine if the tubes were supplied in the spheroidized condition.

As another example, in the case of a bolt failure it is desirable to examine the nuts and other associated parts that may have contributed to the failures. Also, in failures involving corrosion, stress-corrosion, or corrosion fatigue, a sample of the fluid that has been in contact with the metal, or of any deposits that have formed, will often be required for analysis.

Abnormal Conditions. In addition to developing a history of the failed part, it is also advisable to determine if any abnormal conditions prevailed. Determine also whether events—such as an accident—occurred in service that may have initiated the failure, or if any recent repairs or overhauls had been carried out and why. In addition, it is also necessary to inquire whether or not the failure was an isolated example, or if others have occurred, either in the component under consideration or in another of a similar design. In the routine examination of a brittle fracture, it is important to know if, at the time of the accident or failure, the prevailing temperature was low, and/or if some measure of shock loading was involved. When dealing with failures of crankshafts or other shafts, it is generally desirable to ascertain the conditions of the bearings and whether any misalignment existed, either within the machine concerned or between the driving and driven components.

In analysis where multiple components and structures are involved, it is essential that the position of each piece be documented before any of the pieces are touched or moved. Such recording usually requires extensive photography, the preparation of suitable sketches, and the taking and tabulation of appropriate measurements of the pieces.

Next, it may be necessary to take an inventory to determine if all of the pieces or fragments are present at the site of the accident. For example, an investigation of an aircraft accident involves the development of a considerable inventory, including listing the number of engines, flaps, landing gear, and the various parts of the fuselage and wings. It is essential to establish whether all the primary parts of the aircraft were aboard at the time that it crashed. Providing an inventory, although painstaking, is often invaluable. The cause of a complex aircraft accident was determined by an experienced investigator when he observed that a portion of one wing tip was missing from the main impact site. This fragment was subsequently located several miles back on the flight path of the aircraft. The fragment provided evidence of a fatigue failure and was the first component separated from the aircraft, thus accounting for the crash.

The most common problem encountered in examining wreckage involves the establishment of the sequence of fractures so as to determine the origin of the initial failure. Usually, the direction of crack growth can be detected from marks on a fracture surface, such as V-shaped chevron patterns (see Fig. 4(a) and related text). The typical sequence of fractures is shown in Fig. 1, where A and B represent fractures that intersect, and fracture B grew in the direction indicated by the arrow. Here the sequence of fractures is clearly discernible. Obviously, fracture A must have occurred prior to fracture B because the presence of fracture A served to arrest cracking at fracture B. This method of sequencing is called the T-junction procedure and is an important technique in wreckage analysis.

Provided the fragments are not permitted to contact each other, it is also helpful to carefully fit together the fragments of broken components which, when assembled and photographed, may indicate the sequence in which fractures occurred. Figure 2 shows a lug that was part of a pin-joint assembly; failure occurred when the pin broke out of the lug. With the broken pieces of the lug fitted together, it is apparent from the deformation

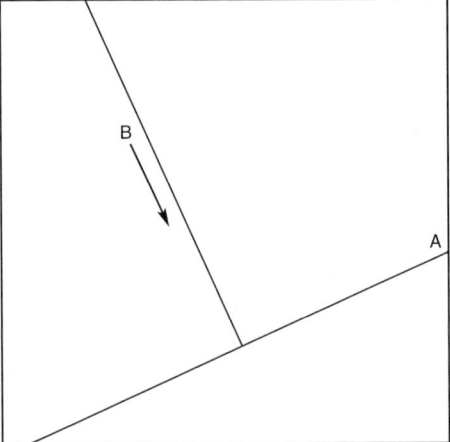

Fig. 1 Schematic of the sequencing of cracking by the T-junction procedure (a technique used in wreckage analysis), where fracture A precedes and arrests fracture B

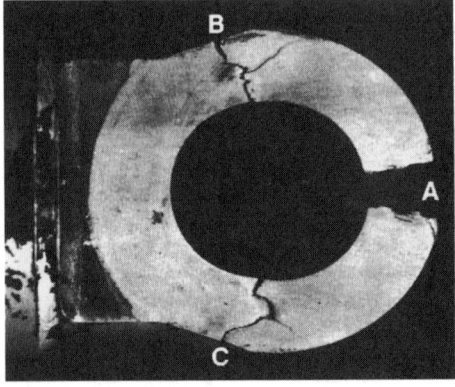

Fig. 2 Fractured lug, part of a pin-joint assembly, showing sequence of fracture. Fracture A preceded fractures B and C

that fracture A must have preceded fractures B and C. However, parts will not fit well together because of plastic deformation that occurred before or during the fracture process.

2. *Preliminary Examination*

The failed part, including all its fragments, should be subjected to a thorough visual examination before any cleaning is undertaken Often, soils and debris found on the part provide useful evidence in establishing the cause of failure or in determining a sequence of events leading to the failure. For example, traces of paint or corrosion found on a portion of a fracture surface may provide evidence that the crack was present in the surface for some time before complete fracture occurred. Such evidence should be recorded photographically.

Visual Inspection. The preliminary examination should begin with unaided visual inspection. The unaided eye has exceptional depth of focus, the ability to examine large areas rapidly and to detect changes of color and texture. Some of these advantages are lost when any optical or electron-optical device is used. Particular attention should be given to the surfaces of fractures and to the paths of cracks. The significance of any indications of abnormal conditions or abuse in service should be observed and assessed, and a general assessment of the basic design and workmanship of the part should also be made. Each important feature, including dimensions, should be recorded, either in writing or by sketches or photographs.

It cannot be emphasized too strongly that the examination should be performed as carefully as possible, because clues to the cause of breakdown often are present but may be missed if the observer is not vigilant. Inspection of the topographic features of the failed component should start with an unaided visual examination and proceed to higher and higher magnification. A magnifying glass followed by a low-power microscope is an invaluable aid in detection of small details of the failed part.

Study of the Fracture. Where fractures are involved, the next step in preliminary examination should be general photography of the entire fractured part, including broken pieces, to record their size and condition and to show how the fracture is related to the components. This should be followed by careful examination of the fracture. The examination should begin with the use of direct lighting and proceed at various angles of oblique lighting to delineate and emphasize fracture characteristics. This should also assist in determining which areas of the fracture are of prime interest and which magnifications will be possible (for a given picture size) to bring out fine details. When this evaluation has been completed, it is appropriate to proceed with photography of the fracture, recording what each photograph shows, its magnification, and how it relates to the other photographs.

For information on photographic equipment, materials, and techniques, the student is referred to the article entitled "Photography of Fractured Parts and Fracture Surfaces," on pages 78 to 90 in *Fractography,* Vol 12 (1987) of the *ASM Handbook.*

3. Nondestructive Inspection

Although often used as quality-control tools, several nondestructive tests are useful in failure investigation and analysis: magnetic-particle inspection of ferrous metals, liquid-penetrant inspection, ultrasonic inspection, and sometimes eddy-current inspection. All these tests are used to detect surface cracks and discontinuities. Radiography is used mainly for internal examination. A photographic record of the results of nondestructive inspection are a necessary part of record keeping in the investigation.

Magnetic-particle inspection utilizes magnetic fields to locate surface and subsurface discontinuities in ferromagnetic materials. When the material or part to be tested is magnetized, discontinuities that generally lie transverse to the direction of the magnetic field will cause a leakage field to be formed at and above the surface of the part. This leakage field, and therefore the presence of the discontinuity, is detected by means of fine ferromagnetic particles applied over the surface, some of which are gathered and held by the leakage field. The magnetically held collection of particles forms an outline of the discontinuity and indicates its size, shape, and extent. Frequently, a fluorescent material is combined with the particles so that discontinuities can be detected visually under ultraviolet light. This method reveals surface cracks that are not visible to the naked eye.

Liquid-Penetrant Inspection. The technique of liquid-penetrant inspection involves the spreading of a liquid penetrant on the sample. This liquid has wetting characteristics, so that it will seep into small cracks and flaws in the sample. The excess liquid is wiped from the surface, and then a developer that causes the liquid to be withdrawn from open cracks or flaws at the surface is applied to the surface. The liquid itself is usually a very bright color or contains fluorescent particles that, under ultraviolet light, reveal discontinuities at the surface of a metal or other material.

Ultrasonic inspection methods depend on sound waves of very high frequency being transmitted through metal and reflected at any boundary such as a metal-air boundary at the surface of the metal, or a metal-crack boundary at a discontinuity within the part or component. High-frequency sound waves can detect small irregularities, but they are easily absorbed, particularly by coarse-grained materials.

The application of ultrasonic testing is limited in failure analysis because accurate interpretations depend on reference standards to isolate the variables. In some instances, ultrasonic testing has proven to be a useful tool in failure analysis, particularly in the investigation of large castings and forgings.

Radiography uses x-rays or gamma rays, which are directed through the sample to a photographic film. After the film has been developed, it can be examined by placing it in front of a light source. The intensity of the light passing through the film will be proportional to the density of the sample and the path length of the radiation. Thus, lighter areas on the plate correspond to the denser areas of the sample, whereas darker areas indicate a crack or defect running in the direction of the incident beam.

The main advantages of radiography are its ability to detect internal discontinuities and to provide permanent photographic records. One disadvantage is that radiography cannot detect cracks whose plane lies perpendicular to the x-ray source.

Residual Stress Analysis. X-ray diffraction is the most common method for direct, nondestructive measurement of residual (internal) stresses in metals. Stresses are determined by measuring the submicroscopic distortion of crystalline lattice structures by tensile or compressive residual stresses. However, it should be pointed out that measurement of residual stresses near fractures or cracks may be erroneous because the residual stresses have already been relieved by the fracture and cracks. Testing of undamaged similar, or exemplar, parts is frequently used as the only alternative in order to understand the residual stress system in the failed part prior to failure.

If the reader is unfamiliar with the basic principles of the various nondestructive methods, reference is made to *Nondestructive Evaluation and Quality Control,* Vol 17 (1989) of the *ASM Handbook,* for detailed descriptions of the methods used.

4. Mechanical Testing

Hardness testing is the simplest of the mechanical tests and is often the most versatile tool available to the failure analyst. Among its many applications, hardness testing can be used to assist in evaluating heat treatment (comparing the hardness of the failed component with that prescribed by specification); to provide an estimate of the tensile strength of steel; to detect work hardening; or to detect softening or hardening caused by overheating, decarburization, or by carbon or nitrogen pickup. Hardness testing is also essentially nondestructive except when preparation of a special hardness test specimen is required, as in microhardness testing. Portable hardness testers are useful for field examination. But the type of hardness test must be appropriate for the sample. For example, Brinell is preferred over Rockwell for a gray cast iron part. One must ensure the proper load is used for the test specimen thickness.

Other mechanical tests are useful in confirming that the failed component conforms to specification or in evaluating the effects of surface conditions on mechanical properties. Where appropriate, tensile and impact tests should be carried out, provided sufficient material for the fabrication of test specimens is available. It may be necessary to make some tests either at slightly elevated or at low temperatures to simulate service conditions. Also, it may be helpful to test specimens after they have been subjected to particular heat treatments simulating those of the failed component in service to determine how this treatment has modified mechanical properties. For example, treating a steel at a temperature in the embrittling range for about one hour prior to impact testing will indicate any tendency to strain-age embrittlement. The determination of the ductile-brittle transition temperature may be useful in investigating brittle fracture of a low-carbon steel.

The failure analyst should exercise care in interpreting mechanical test results. If a material has a tensile strength 5 to 10% below the minimum specified, this does not mean that low hardness or strength is the cause of its failure in service. Also, it should be understood that laboratory tests on small specimens may not adequately represent the behavior of a much larger structure or component in service. For instance, it is possible for a brittle fracture of a large structure to occur at or near ambient temperature, while subsequent laboratory tests of Charpy or Izod specimens show a transition temperature well below -18 °C (0 °F). The effects of size in fatigue, stress-corrosion, and hydrogen embrittlement testing are not well understood. However, on the basis of the limited evidence available, it appears that resistance to these failure processes decreases as specimen size increases. Several investigators have found correlation problems of transition-temperature-type impact tests with service performance.

Tensile tests, in many failure analysis investigations, do not provide enough useful information because relatively few failures result from metal that is deficient in tensile strength. Furthermore, samples cut from components that have failed in a brittle manner generally show adequate ductility under the conditions imposed during a tensile test.

Sometimes, however, there is justification for tensile testing of failed components to eliminate poor-quality material as a possible cause of failure. Often, these tensile tests for determining material quality are carried out by manufacturers and suppliers when examining components that have been returned to them for analysis.

The role of directionality in tensile testing of wrought metals should also be considered. Specimens cut transversely to the longitudinal axis of a component (such as a shaft, plate, or sheet) usually give lower tensile and ductility values than those cut along the longitudinal axis. This is due to the marked directionality and the resulting anisotropy produced during rolling or forging.

5. *Selection and Preservation of Fracture Surfaces*

The proper selection, preservation, and cleaning of fracture surfaces is vital to prevent important evidence from being destroyed or obscured.

Surfaces of fractures may suffer either mechanical or chemical damage. Mechanical damage may arise from several sources, including the striking of the surface of the fracture by other objects. This can occur during actual fracture in service or when removing or transporting a fractured part for analysis.

Usually, the surface of a fracture can be protected during shipment by a cloth or cotton covering, but this may remove some loosely adhering material, which might contain the primary clue to the cause of the fracture. Touching or rubbing the surface of a fracture with the fingers should definitely be avoided. Also, no attempt should be made to fit together the sections of a fractured part by placing them in contact. This generally accomplishes nothing and almost always causes damage to the fracture surface. The use of corrosion inhibitor paper to package samples should be considered.

Chemical (corrosion) damage to a fracture specimen can be prevented in several ways. For instance, because the identification of foreign material present on a fracture surface may be important in the overall determination of the cause of the fracture, many laboratories prefer not to use corrosion-preventive coatings on a fracture specimen. When possible, it is best to dry the fracture specimen, preferably by using a jet of dry, compressed air (which will also blow extraneous foreign material from the surface), and then to place it in a desiccator or pack it with a suitable desiccant. However, clean, fresh fracture surfaces should be coated when they cannot be protected from the elements. Several days may be required to remove critical specimens from large structures so that coating the fracture surfaces would be the prudent decision.

Cleaning of fractured surfaces should be avoided in general, but must be done for examination with a scanning electron microscope (SEM) and often to reveal macroscale fractographic features. Cleaning should proceed in stages using the least aggressive procedure first, then proceeding to more aggressive procedures if needed. Washing the fracture surface with water should especially be avoided. However, specimens contaminated with sea-water or with fire-extinguishing fluids require thorough washing, usually with water, followed by a rinse with acetone or alcohol before storage in a desiccator or coating with a desiccant. Sometimes cleaning may also be required for removal of obliterating debris and dirt or to prepare the fracture surface for electron microscope examination. Other acceptable cleaning procedures include use of a dry-air blast or of a soft-hair artist's brush; treating with inorganic solvents, either by immersion or by jet; treating with mild acid or alkaline solutions (depending on the metal) that will attack deposits but to which the base metal is essentially inert; ultrasonic cleaning; and application and stripping of plastic replicas.

Cleaning with cellulose acetate tape is one of the most widely used methods, particularly when the surface of a fracture has been affected by corrosion. A strip of acetate about 0.1 mm (0.005 in.) thick and of suitable size is softened by immersion in acetone and placed on the fracture surface. The initial strip is backed by a piece of unsoftened acetate, and then the replica is pressed hard onto the surface of the fracture using a finger. The drying time will depend on the extent to which the replicating material was softened, and this in turn will be governed by the texture of the surface of the fracture. Drying times of at least 15 to 30 min are recommended. The dry replica is lifted from the fracture, using a scalpel or tweezers. The replicating procedure must be repeated several times if the fracture is badly contaminated. When a clean and uncontaminated replica is obtained, the process is complete. An advantage of this method is that the debris removed from the fracture is preserved for any subsequent examination that may be necessary for identification by x-ray or electron diffraction techniques. To be complete, the analyst should filter solvents used for cleaning to recapture insoluble particulates.

Sectioning. Because examination tools, including hardness testers and optical and electron microscopes, are limited as to the size of specimen they can accept, it is often necessary to remove from a failed component a fracture-containing portion or section that is of a size convenient to handle and examine. The student is reminded that this is a destructive process and should be aware of the spoliation of evidence in potential litigation cases.

Before cutting or sectioning, the fracture area should be carefully protected. All cutting should be done so that surfaces of fractures and areas adjacent to them are not damaged or altered; this includes keeping the fracture surface dry, whenever possible. For large parts, the common method of removing specimens is by flame cutting. Cutting must be done at a sufficient distance from the fracture site so that the microstructure of the metal underlying the surface of the fracture is not altered by the heat of the flame, and so that none of the molten metal from flame cutting is deposited on the surface of the fracture.

Heat from any source also can affect metal properties and microstructures during cutting. Therefore, dry abrasive cut-off wheels should never be used near critical surfaces that will be examined microscopically. Therefore, sectioning should be performed with jewelers' saws, precision diamond-edged, thin cut-off wheels, hacksaws, bandsaws, or soft abrasive cut-off wheels flooded with water-based soluble oil solution to keep metal surfaces cool and corrosion-free. Dry cutting with an air-driven abrasive disk may also be used with care to remove small specimens from large parts if kept cool along with coating the fracture surface for protection.

Secondary Cracks. When the primary fracture has been damaged or corroded to such a degree that most of the information relevant to the cause of the failure is obliterated, it is desirable to open any secondary cracks to expose their fracture surfaces for examination and study. These cracks may provide more information than the primary fracture.

In opening cracks for examination, care must be exercised to prevent damage, primarily mechanical, to the surface of the fracture. This can usually be accomplished if opening is done in such a way that the two surfaces of the fracture are moved in opposite directions, normal to the fracture plane. Generally, a saw cut can be made from the back of the fractured part to a point near the tip of the crack, using extreme care to avoid actually reaching the tip of the crack. This saw cut will reduce the amount of solid metal that must be broken. The final breaking of the specimen can be done in several ways: (a) by clamping the two sides of the fractured part in a tensile-testing machine, if the shape permits, and pulling, (b) by placing the specimen in a vise and bending one half away from the other by striking it with a hammer in a manner that will avoid damage to the surfaces of the crack, or (c) by gripping the halves of the fracture in pliers or vise grips and bending or pulling them apart. Cooling the part with liquid nitrogen often reduces the force and plastic deformation necessary to fracture the part. Fortunately there is little confusion during subsequent examination as to which part of the fracture surface was obtained in opening the crack.

It is recommended that both the crack separation and the visible crack length be measured prior to opening. The analyst may have to use dye penetrant or other nondestructive evaluation (NDE) technique to actually see the length of a tightly closed crack. Often, the amount of strain that occurred in the specimen can be determined from a measurement of the separation between the adjacent halves of a fracture. This should be done before preparation for opening a secondary crack has begun. The lengths of cracks may also be important for analyses of fatigue fractures or for consideration for the application of fracture mechanics.

6. *Macroscopic Examination of Fracture Surfaces*

One very important part of any failure analysis is the macroscopic examination of fracture surfaces. Performed at magnifications from 1 to 50 or 100 times, it may be conducted by the unaided eye, a hand lens or magnifier, a low-power stereoscopic microscope, or an SEM. Macroscopic photography of up to 20 times magnification requires a high-quality camera and special lenses; alternatively, a large magnifying glass may be used to enlarge a specific area in the photo, such as a crack or other small detail. A metallographic microscope with macro-objectives and lights may be used for somewhat higher magnifications. However, depth-of-field becomes extremely limited with light optics. For much greater depth-of-field, an SEM may be used for low-magnification photography as well as higher magnification work. Stereo or three-dimensional photographs may also be made to reveal the topographic features of a fracture or other surface.

Frequently, a specimen may be too large or too heavy for the stage of the metallograph or the chamber of an SEM, and cutting or sectioning the specimen may be difficult or not allowed because of legal limitations at the time. In these instances, excellent results can be achieved by examining and—where appropriate—photographing replicas made by the method for cleaning fractures (see discussion under Cleaning). These replicas can be coated with a thin layer (about 20 nm, or 2×10^{-8} m, thick) of vacuum-deposited gold or aluminum to improve their reflectivity, or they may be shadowed at an angle to increase the contrast of fine detail. The replicas may be examined by incident-light or transmitted-light microscopy. Because they are electrically conductive, the coated replicas may also be examined by SEM.

The amount of information that can be obtained from examination of a fracture surface at low-power magnification is extensive. The orientation of the fracture surfaces must be consistent with the proposed mode of failure and the known loads on the failed part. Failure in monotonic tension produces a flat (square) fracture normal (perpendicular) to the maximum tensile stress and frequently a slant (shear) fracture at about 45°. This 45° slant fracture is often called a "shear lip." Many fractures are flat at the center, but surrounded by a "picture frame" of slant fracture. An example of this behavior is to be found in the familiar cup-and-cone fracture of a round tensile test bar.

In thin sheets, tube walls, and small diameter rods, slant shear fracture may occur because through-the-thickness stresses are minimized—that is, even though there may be a plane strain condition, there are minimal hydrostatic tensile stresses.

Macroscopic examination can sometimes reveal the direction of crack growth and hence the origin of failure. With brittle, flat fractures, determination depends largely on the fracture surface showing chevron marks, a radial fanlike pattern, of the type shown in Fig. 3. Cracks propagate

Fig. 3 Origin (at arrow) of a single-load brittle fracture that initiated at a small weld defect. Note also a fatigue fracture in the upper right corner. Radial ridges emanate from the origin in a fan-shaped pattern.

parallel to shear lips if they are present. Where fracture surfaces show both flat and slant surfaces, this can be the terminal end of a fast-moving brittle fracture where the crack speed has slowed significantly. Crack extension can relax the stress so that final fracture occurs by slant shear fracture. Conversely, if a fracture has begun at a free surface, the fracture-origin area is usually characterized by a total absence of slant fracture or shear lip.

Low-power examinations of fracture surfaces often reveal regions having a texture different from the region of final fracture. Fatigue, stress-corrosion, and hydrogen embrittlement fractures may also show these differences because the final failure is due to overload after the cross section is reduced by one of the aforementioned crack initiation modes.

Figure 4(a) shows the fracture surface of a steel tube and is an excellent example of the type of information that can be obtained by macroscopic examination. The V-shaped chevron marks and fanlike marks clearly indicate that the fracture origin is at the point marked by the arrow. This region, unlike the rest of the fracture, has no shear lip. The flat fracture surface suggests that the stress causing the failure was tension parallel to the length of the tube. The origin of the fracture as seen at higher magnification in Fig. 4(b) shows several small fracture origins having a texture different from that of the remainder of the fracture surface.

Figure 5 shows the fatigue fracture surface of a broken shaft. As indicated, the fracture originated at the corner of the keyway of this shaft, and the beach marks swing counterclockwise due to the clockwise rotation of the shaft.

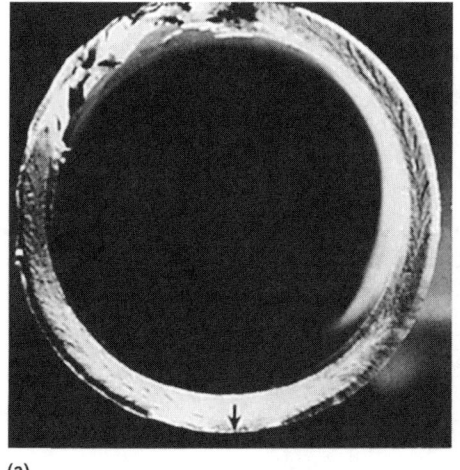

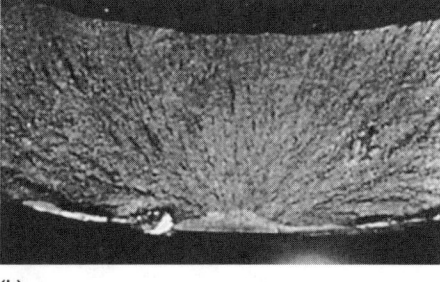

(a) (b)

Fig. 4 (a) Fracture surface of a steel tube, at approximately actual size, showing point of crack initiation (at arrow), chevron and fanlike marks, and development of shear lips. (b) Fracture-origin area. Original magnification 5×; note that fracture nuclei differ in texture from the main fracture surface

166 / How to Organize and Run a Failure Investigation

7. Microscopic Examination of Fracture Surfaces

Microscopic examination of fracture surfaces is typically done with an SEM. An SEM has the advantage over light microscopy because of the large depth of field and very high magnifications attainable, typically 5000 to 10,000×. In addition, SEMs are often equipped with microanalytical capabilities, for example, energy-dispersive x-ray (EDX) spectroscopes. Chemical analysis can be helpful in confirming the chemistry of microstructural features that may be confused with fracture features.

The primary limitation of SEM analysis is sample size. Scanning electron microscopic analysis must be conducted in a vacuum so the sample must be put into a chamber that typically holds a sample less than 205 mm (8 in.) in diameter. Although there are some fracture surface features that are commonly associated with particular failure modes, the novice failure analyst must be very careful in fractographic analyses. The features listed below are some of the more classic examples of fracture surface topography that indicates a fracture mode:

- Dimpled rupture typical of overstress failures of ductile metals and alloys (see Fig. 6)
- Cleavage facets, typical of transgranular brittle fracture of body-centered cubic (bcc) and hexagonal close-packed (hcp) metals and alloys (see Fig. 7)
- Brittle intergranular fracture typical of temper-embrittled steel, where fracture is due to segregation of an embrittling species to grain bound-

Fig. 5 The offsetting effect of rotation on fatigue fracture beach marks reveals the direction of shaft rotation during operations

aries (such as oxygen in iron or nickel), due to intergranular stress-corrosion cracking (IGSCC) or due to hydrogen embrittlement (see Fig. 8)
- Stage II striations, typical of fatigue failure (see Fig. 9)

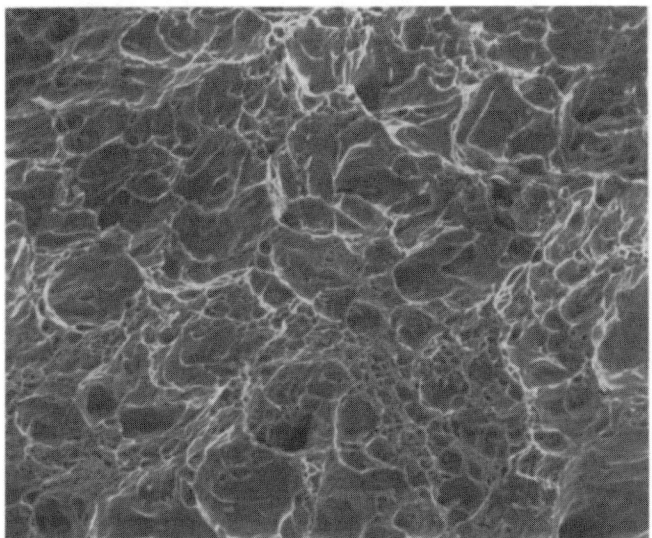

Fig. 6 Typical dimpled rupture fracture surface of a ductile fracture viewed at a magnification of 2000× and at an angle of about 40–50° to the fracture surface

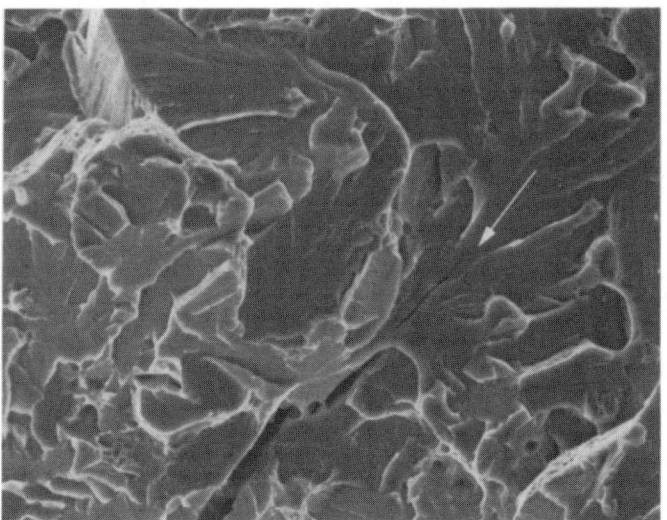

Fig. 7 Cleavage fracture in hardened steel, viewed under the scanning electron microscope. Note progression of "river" marks in the direction of arrow. Grain boundaries were crossed without apparent effect. Original magnification at 2000×

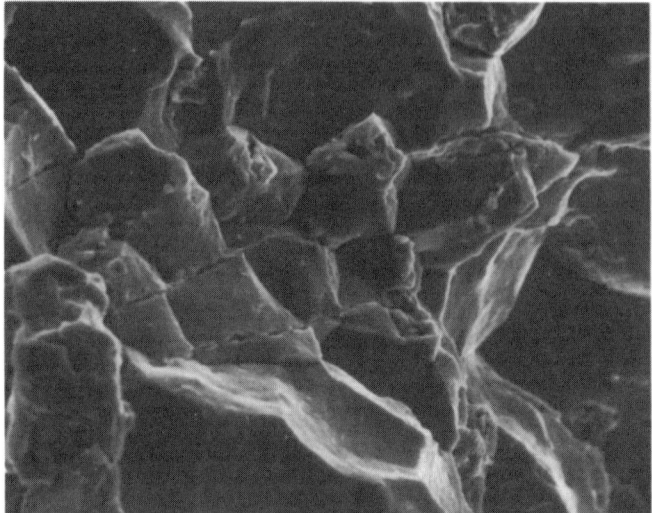

Fig. 8 Intergranular fracture in hardened steel, viewed under the scanning electron microscope. Note that fracture takes place between the grains; thus the fracture surface has a "rock-candy" appearance that reveals the shapes of part of the individual grains. Original magnification at 2000×

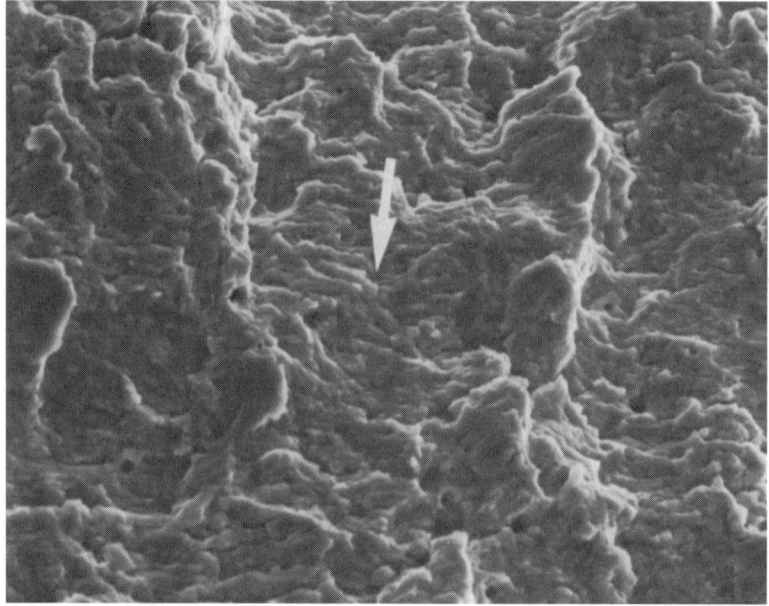

Fig. 9 Fatigue striations in low-carbon alloy steel (8620). This scanning electron microscope fractograph shows the roughly horizontal ridges, which are the advance of the crack front with each load application. The crack progresses in the direction of the arrow. Original magnification at 2000×

These fracture surface features are described in greater detail in *Principles of Failure Analysis* (Education Course), Lesson 3, ASM International, 2002, along with other important aspects of fractography

8. Metallographic Examination

Metallographic examination of polished and of polished-and-etched sections by optical microscopy and by electron-optical techniques is a vital part of failure investigation and should be carried out as a routine procedure when possible. Metallographic examination provides the investigator with a good indication of the class of material involved and its structure. If abnormalities are present, these may be associated with undesirable characteristics that predispose to early failure. It is sometimes possible to relate them to an unsuitable composition or to the effects of service, such as aging in low-carbon steel that has caused precipitation of iron nitride, or gassing in copper. Microstructural examination may also provide information as to the method of manufacture of the part under investigation. It can reveal the heat treatment and possible deficiencies in heat treatment such as decarburization at the surface. Microstructural inspection can also reveal possible overheating through coarsening of carbides of superalloys and solution and precipitation of manganese sulfide. Other service effects, such as corrosion, oxidation, and severe work hardening of surfaces also are revealed, and their extent can be investigated. The topographical characteristics of any cracks, particularly their mode of propagation, can be determined, for example, transgranular or intergranular. This provides information that can be helpful in distinguishing between different modes of failure. For example, fatigue cracks always propagate perpendicular to the maximum cyclic tensile stress while stress-corrosion cracking (SCC) may propagate along grain boundaries.

Only a few general directions can be given as to the best location from which to take specimens for microscopic examination, because almost every failure has individual features to be taken into account. In most examinations, however, it must be determined whether the structure of a specimen taken adjacent to a fracture surface or a region at which a service defect has developed is representative of the component as a whole. This can be done only by the examination of specimens taken from the failure region and specimens taken from other locations. For instance, in the case of ruptured or bulged boiler tubes in which failure is usually restricted to one portion only, it is desirable to examine specimens taken from both sides of the fracture, from a location opposite the affected zone, and also from an area as remote from the failure as the size of the sample permits, so as to determine whether the failure has been due to a material defect or to overheating—and, if the latter, whether this was of a general or localized nature. In investigations involving general overheating, sometimes the original condition of the material can be ascertained only from a sample cut from a part of the tube many feet away from the affected zone.

Metallographic specimens should be taken perpendicular to the fracture surface, showing the fracture surface in edge view. In cases where metal cleanliness may be an issue, the specimen orientation must be selected properly to determine inclusion density and morphology. This type of examination must be performed on the unetched metallographic specimen.

In the investigation of fatigue cracks, it may be desirable to take a specimen from the region where the fracture originated to ascertain if the initial development was associated with an abnormality, such as a weld defect, a decarburized surface, a zone rich in inclusions, or in castings a zone containing severe porosity. Multiple fatigue-crack initiation is very typical of both fretting and corrosion fatigue and may form in areas where there is constant stress across a section. Figure 10 shows an example of a large piston rod fracture from a forging hammer. The fracture originated at the surface due to severe friction during contact with the interior of the hole in the upper die block.

Similarly, with surface marks, where the origin cannot be identified from outward appearances, a microscopic examination will show whether they occurred in rolling or arose from ingot defects, such as scabs, laps,

(a)   (b)

Fig. 10 An example of a large piston rod fracture from a forging hammer. The fracture originated at the surface due to severe friction during contact with the interior of the hole in the upper die block. (a) In addition to the main fracture there are many fatigue cracks in the darker areas. (b) Note the smooth surface near the origin of the fracture because of the severe pounding during service. Note the ratchet marks separate the fatigue areas. The shear lip, top, was the last region to fracture.

or seams. In brittle fractures, it is useful to examine a specimen cut from where the failure originated, if this can be located with certainty. Failures by brittle fracture may be associated with locally work-hardened surfaces, arc strikes, local untempered martensite, and so forth.

For good edge retention when looking at a fracture surface, it is usually best to plate the surface of a specimen with a metal, such as nickel, prior to mounting and sectioning, so that the fracture edge is supported during grinding and polishing and can be included in the examination. Alternative means include hard metal or nonmetal particles embedded in the mount adjacent to the edges.

Analysis of Metallographic Sections. As with hardness testing and macroscopic examination, the examination of metallographic sections with a microscope is standard practice in most failure analyses, because of the outstanding capability of the microscope to reveal material imperfections caused during processing and of detecting the results of a variety of in-service operating conditions and environments that may have contributed to failure. Inclusions, microstructural segregation, decarburization, carbon pickup, improper heat treatment, untempered white martensite, dissolved second phases such as gamma prime in nickel-base superalloys and intergranular corrosion are among the many metallurgical imperfections and undesirable conditions that can be detected and analyzed by microscopic examination of metallographic sections.

Figures 11 and 12 illustrate the usefulness of metallographic sectioning in failure analysis.

Figure 11 shows a longitudinal section of a fracture in a driveshaft propagating from a built-up weld.

Figure 12 shows a carburized steel gear that failed by both plastic flow and destructive pitting An error in machining caused an excessive amount

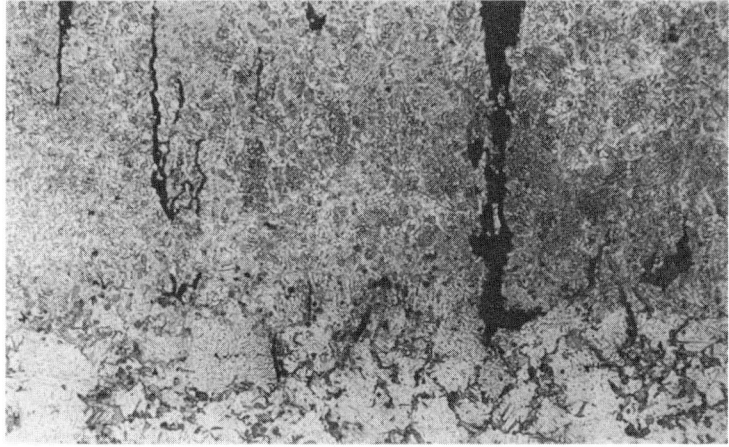

Fig. 11 Fracture of driveshaft propagating from built-up weld. The microstructure shows transition from weld (top) to base metal (bottom). Longitudinal section. Nital etched. Original magnification at 100×

of stock removal from the drive faces of the teeth. Stock removal caused a reduced case depth and therefore the shallow case was incapable of supporting the necessary loads. The drive faces of the gear teeth were deformed by plastic flow and severely pitted.

Even in the absence of a specific metallurgical imperfection, examination of metallographic sections is invaluable to the investigator in the measurement of microstructural parameters such as case depth, grain size, thickness of plated coatings, and heat-affected zone size—all of which may have a bearing on the cause of failure. For the analyst less experienced in microstructural examinations, *Metallography and Microstructures,* Vol 9 (2004) of the *ASM Handbook,* is a valuable resource for identifying unknown structures.

9. Stress Analysis

It is sometimes quite apparent that an excessively high load or stress level was the direct cause or contributed significantly to the failure. Even so, an accurate stress analysis of the magnitude and type (axial, torsion, bending) of stress is required to substantiate the role of stress. In other failure analyses, the analyst may have strong evidence that the cause of a

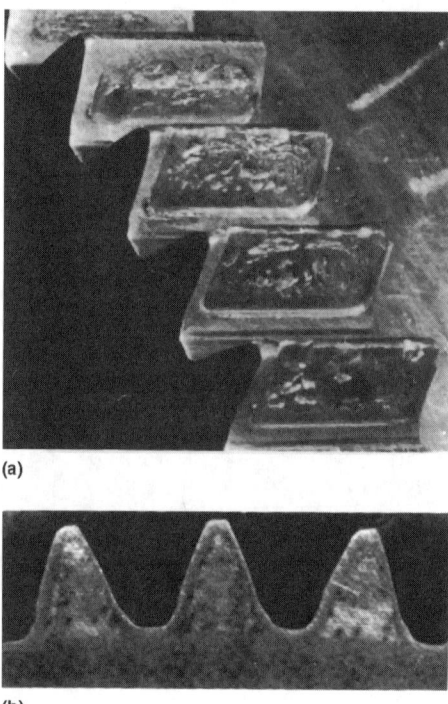

Fig. 12 (a) Damage involving both plastic flow and destructive pitting on teeth of a carburized AMS 6260 steel gear. (b) Etched end face of the gear, showing excessive stock removal from drive faces of teeth

failure is related to excessively high static stresses (or cyclic stresses in the case of fatigue). In these cases, an analysis of the stress during normal operation (or abnormal operation if identified) must be conducted. Analytical, closed-form calculations based on engineering mechanics are often used by designers to predict stress levels in the early design stages. This method of using known "machine design and structural" formulas to predict the stress under a given load is also helpful to the failure analyst especially in cases where this step may not have been used in the original design of the part. It is not uncommon to find products that have no record of any stress analysis in the development of the product. Even if such calculations were made, the failure analyst may not have access to them. The analyst must answer the questions "Was the component sized properly by the design stress analysis? Did the material have the properties assumed in the design? Did the part fail in a manner consistent with that assumed in design, or did it fail in a way not anticipated in the original design?"

In cases where unusual or abnormal loading is suspected, direct calculation of stresses will fall short and predict incorrect stress levels. In these cases, experimental stress analysis is used for determining machine loads and component stresses. This technique normally involves attachment of strain gages to similar parts in critical areas or typical areas where the failure has occurred. The strain gages are connected to a monitoring device either directly with wires or indirectly by radio signals for monitoring moving or rotating parts. In this way the actual, dynamic stresses can be determined.

For products with very complex shapes and high thermal gradients, a finite-element analysis (FEA) may be required to estimate the level of stress that most likely existed in the failed component. These analyses can stand alone or can be used to help select critical locations for strain gage attachment. Finite-element analyses can be time consuming and expensive, but they are necessary for an accurate assessment of stress levels in areas of complex geometry of some components. This type of analysis is almost essential for determining stresses caused by thermal gradients such as those found in welding.

Overload failures are often a result of improper design or improper operation. A design analysis is essential in determining which of these is the root cause. Sometimes the improper design is a result of incorrect information passed to the designer. In these cases, a failure is the only indication that the wrong inputs were used for the design. This is also true for fatigue failures. For proper design of rotating or moving parts, a detailed stress analysis is essential. It is much more difficult to predict dynamic stresses than static stresses.

10. Failure Modes

Because the initial steps in failure analysis of a fracture involve visual and macroscopic observation, the first impressions should be based on obvious visual evidence. The simplest and most important observations relate to deformation: Was the metal obviously deformed?

If it was deformed prior to fracture, yielding and fracture has occurred due to one or more gross overloads. It is predominantly a ductile fracture or a very high-stress, low-cycle fatigue fracture, as can be demonstrated by repeated manual bending of a paper clip or wire coat hanger. The deformation is directly related to the type of stress causing fracture: tension (stretched), bending (bent), torsion (twisted), or compression (shortened or buckled), or a combination of these stress types.

The absence of gross deformation of the failed part indicates that the fracture is predominantly brittle. A brittle fracture should not be confused with brittle material. The shape or geometry of the part made from a ductile metal can result in an overload failure with little overall shape change, or a failure mechanism can operate to start and grow a crack such as a fatigue crack or stress-corrosion crack. When such a crack grows to the point that the remaining cross-sectional area of the part is overloaded by the normal loads, the final overload failure has little macroscale deformation associated with it. Thus, on a macroscale, the failure of a ductile metal can appear brittle. Of course, overload failures of brittle material always appear brittle on a macroscale.

It is usually more difficult to analyze a brittle fracture because there are a large number of possible mechanisms that can cause fracture with little or no obvious deformation. For single overload fractures, these include such factors as stress concentrations, low temperatures, high rates of loading, high metal strength and hardness, SCC, hydrogen embrittlement, temper embrittlement, large section size, and others. For fatigue fractures, causative factors can include stress concentrations, tensile residual stresses, large stress amplitudes, large numbers of load applications, corrosive environments, high temperatures, low metal strength and hardness, wear, and others.

From this discussion, it should become clear that proper failure analysis is not simple, but can become exceedingly complex, requiring considerable thought, examination, questioning, and reference to other sources of information in the literature, as suggested at the end of this appendix. However, identifying the failure mode is the key step in a failure analysis, and it is the essential part of determining the root cause.

Ductile Fracture. Overload fractures of many metals and alloys occur by ductile fracture. Overloading in tension is perhaps the least complex of the overload fractures, although essentially the same processes operate in bending and torsion as well as under the complex states of stress that may have produced a given service failure.

The classic example of ductile failure is a tensile test. In this fracture process, considerable elongation—that is, deformation—takes place before the geometric instability, necking, begins. Even after the deformation is localized at the neck, significant deformation occurs at the neck before the fracture process begins. After the neck forms, the curvature of the neck creates a triaxial tensile stress state. This leads to initiation of an internal crack near the center of the necked region. In commercial grade alloys, discontinuities such as inclusions or second-phase particles are sources of early void formation by separation of the matrix and the particle. Some of these voids coalesce to develop a crack, which is perpendicular to the tensile axis. The crack spreads until the state of stress, ductility of the metal, and flow condition reach a condition that favors a shear displacement. The crack path then shifts to a maximum shear plane, which is at an angle to the tensile axis (close to 45° in cylindrical specimens). This portion of the fracture surface is known as the shear lip. Sometimes this shear lip forms only on one side of the initial flat crack. When this occurs, the resulting fracture surface has a macroscopic appearance known as cup-and-cone. For brittle materials, the majority of the fracture surface is perpendicular to the tensile axis with little or no fracture surface lying on a plane of shear.

Ductile failures in biaxially loaded sheet and plate structures often consist entirely of a shear lip. Pipe and pressure vessels are examples of biaxially stressed components. Often, failures in these components may first appear brittle with limited ductility. However, close inspection usually reveals some general thickness reduction but no necking at the fracture surface.

Features of Ductile Fracture. High-magnification examination of ductile fracture surfaces usually reveals approximately equiaxed dimples (see Fig. 6), sometimes the particles that originated the individual dimples are visible. Slant fractures or ductile fracture on planes of high shear stress generate elongated dimples. When the elongated dimples are produced by a shear component, the dimples in the mating fracture surfaces point in opposite directions. Some investigators have described ductile fracture surfaces produced by tearing and suggest that the crack produces elongated dimples on mating surfaces that are mirror images. This model has recently been challenged. Few if any observations of crack propagation causing elongated dimples have been reported.

Ductile fractures are usually transgranular, but electron fractography examinations have shown dimple patterns on tensile-fracture surfaces of aluminum-copper alloys when metallographic sections have shown that the fracture path was apparently intergranular and nominally brittle. Isolated dimples on brittle intergranular-fracture surfaces of Al-Zn-Mg alloys have been observed. These observations are generally confined to precipitation-hardened alloys with grain-boundary zones that are precipitate-free.

Brittle Fracture. There are two general types of brittle fracture caused by a single overload: transgranular cleavage and intergranular separation. Each has distinct features that make identification relatively simple.

Transgranular cleavage (through the grains) of iron and low-carbon steel is the most commonly encountered process of brittle fracture—so common, in fact, that the term brittle fracture is sometimes misinterpreted as meaning only transgranular cleavage of iron and low-carbon steel. Transgranular cleavage can also occur in several other bcc metals and their alloys (for example, tungsten, molybdenum, and chromium) and some hcp metals (for example, zinc, magnesium, and beryllium). Face-centered cubic (fcc) metals and alloys (such as aluminum and austenitic stainless steels) are usually regarded as immune from this fracture mechanism.

Iron and low-carbon steels show a ductile-to-brittle transition with decreasing temperature that arises from a strong dependence of the yield stress on temperature. Brittle fracture of normally ductile metals depends on several physical factors, including specimen shape and size, temperature, and strain rate. Thus, a component or structure that has given satisfactory service may fracture unexpectedly; the catastrophic brittle fracture of ships in heavy seas and the failure of bridges on unusually cold days are examples. Metallurgical changes, especially strain aging, may cause the brittle fracture of such items as crane hooks and chain links after long periods of satisfactory operation.

Cleavage fracture is not difficult to diagnose; the fracture path is by definition crystallographic. In polycrystalline specimens, this often produces a pattern of brightly reflecting crystal facets, and such fractures are sometimes described as crystalline. The general plane of fracture is approximately normal (perpendicular) to the axis of maximum tensile stress, and a shear lip is often present as a "picture frame" around the fracture. The local absence of a shear lip or slant fracture suggests a possible location for initiation of the fracture, for shear lips form during the final stages of the fracture process.

The fractography of cleavage fracture in low-carbon steels, iron, and other single-phase, bcc metals and alloys is fairly well established. In polycrystalline specimens, numerous fan-shaped cleavage plateaus, usually showing a high degree of geometric perfection, are present (see Fig. 7). The most characteristic feature of these plateaus is the presence of a pattern of river marks, which consist of cleavage steps or tear ridges and indicate the local direction of crack growth. The rule is that, if the tributaries are regarded as flowing into the main stream, then the direction of crack growth is downstream. This is in contrast to macroscopic chevron marks, where the direction of crack growth, using the river analogy, would be upstream.

Other fractographic features that may be observed include the presence of cleavage on conjugate planes, tear ridges, ductile tears joining cleavage planes at different levels and tongues, which result from fracture in mechanical twins formed ahead of the advancing crack. Cleavage fracture in pearlitic and martensitic steels is less easily interpreted, because microstructure tends to modify the fracture surface. In fact, cleavage fracture surfaces of pearlitic steel has characteristics similar to fatigue striations, so one must be careful not to confuse the fracture mode.

Intergranular (between the grains) fracture can usually be recognized, but determining the primary cause of the fracture may be difficult. Fractographic and microscopic examination can readily identify the presence of second-phase particles at grain boundaries. Unfortunately, the segregation of a layer a few atoms thick of some element or compound that produces intergranular fracture often cannot be detected by fractography. Auger analysis and sometimes energy-dispersive spectroscopy (EDS) are useful for very thin layers. Some causes of intergranular brittle fracture are given below, but the list is not exhaustive. It does, however, indicate some of the possibilities that need to be considered, and either eliminated or confirmed, as contributing to the fracture.

- The presence at a grain boundary of a large area of second-phase particles (such as carbides in Fe-Ni-Cr alloys or manganese sulfide, or MnS, particles in an overheated steel).
- Segregation of a specific element or compound to a grain boundary where a layer a few atoms thick is sufficient to cause embrittlement. Embrittlement caused by the presence of oxygen in high-purity iron, oxygen in nickel, or antimony in copper, and temper embrittlement of certain steels are examples of intergranular embrittlement where detection of a second phase at grain boundaries is difficult.

The conditions under which a progressively growing crack may follow an intergranular path before final fracture occurs include SCC, embrittlement by liquid metals, hydrogen embrittlement, as well as creep and stress-rupture failures. These failure modes are discussed in the paragraphs that follow.

Fatigue (repetitive load) fracture results from the application of repeated or cyclic stresses, each of which may be substantially below the nominal yield strength of the material. There are a great many variables that influence fatigue behavior; these include the magnitude and frequency of the fluctuating stress, the presence of a mean stress, temperature, environment, specimen size and shape, state of stress, the presence of residual stresses, surface finish, microstructure, and the presence of fretting damage. An additional problem is that one variable may be more important to one material than another. For example, high-strength materials are more sensitive to notches or stress concentrations than lower-strength materials.

The most noticeable macroscopic features of classic fatigue-fracture surfaces are the progression marks (also known as beach marks, clamshell marks, or tide marks) that indicate successive positions of the advancing crack front (see Fig. 13). As opposed to actual microscopic striations that are individual steps of the advancing fatigue crack, macroscopic beach marks are generally slight deviations in the direction of the fatigue crack.

Fatigue-fracture surfaces are smooth and textured (or rubbed) near their origins and generally show slight roughening as the crack grows. There is little or no macroscopic deformation associated with the fatigue portion of the crack growth prior to overload fracture, and there may be some evidence that the crack has followed specific crystal planes during early growth, thus giving a faceted appearance.

However, once a crack develops it is driven by tensile forces and remains perpendicular to the principal tensile axis. Unfortunately, a great many fatigue fractures do not show the classic progression marks, frequently because of inadequate microscopic ductility (too high hardness) or because of a significant quantity of brittle or weak second-phase particles (such as graphite in cast iron).

Most fatigue cracks are transgranular without marked branching, although intergranular fatigue can occur. Corrosion fatigue in most materials is also transgranular; its most striking feature usually is the multiplicity of crack origins, only one of which extends catastrophically. Fatigue initiated by fretting (rubbing) has similar characteristics and is generally diagnosed by the presence of a corrosion product filling the multiplicity of cracks and by the presence of fretting debris on the surface of the component. For aluminum alloys, the fretting product is often a hard, black deposit; on steels, considerable quantity of a rustlike material is produced. The fretting product appears to be a mixture of finely divided particles of the base metal, its oxides, and its hydrated oxides. The role of fretting in fatigue is the dramatic reduction of the endurance limit by altering the surface condition. Fretting damage has been blamed for as much as a 75% reduction of the endurance limit for machined steel.

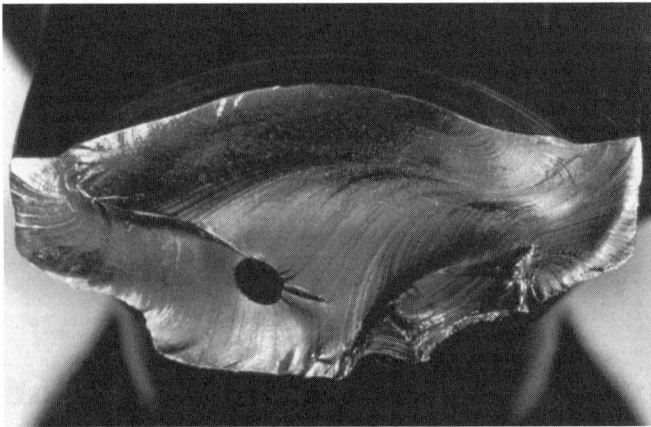

Fig. 13 A classic fatigue-fracture surface showing progression marks (beach marks) that indicate successive positions of the advancing crack front

Microscopically, surfaces of fatigue fractures are frequently characterized by the presence of striations, each of which is produced by a single cycle of stress. Figure 9 shows classical fatigue striations in low alloy steel. However, every stress cycle does not necessarily produce a striation; in fact, the absence of striations does not rule out fatigue, as mentioned above for cast iron. Also, there are a number of fractographic features that may be confused with fatigue striations—Wallner lines, produced by shock-wave/crack-front interactions, rub marks, microstructural features such as pearlite, and various structures in nonferrous alloys.

Stress-corrosion cracking (SCC) is a mechanical/environmental failure process in which mechanical stress and chemical attack combine to cause the initiation and propagation of fracture in a metal part. It is produced by the synergistic action of a sustained tensile stress and a specific corrosive environment, causing failure in less time than would the sum of the separate effects of the stress and the corrosive environment.

Failure by SCC is frequently caused by exposure to a seemingly mild chemical environment while subject to an applied or residual tensile stress that may be well below the yield strength of the metal. Under such conditions, fine cracks can penetrate deeply into the part, although the surface may show only insignificant amounts of surface discoloration or pitting. Hence, there may be no macroscopic indications of impending failure. The most common instances of failure by SCC in service are probably those associated with the following metals and alloys:

- High-strength aluminum alloys, especially of the Al-Zn-Mg type, under atmospheric-corrosion conditions; internal and assembly stresses are often important
- Austenitic stainless steels and nickel alloys of the Inconel type in the presence of very low concentrations of chloride ions; the hydroxyl ion is also reported as causing SCC
- Low-carbon structural steels, usually in the presence of hot concentrated nitrate or caustic solutions, or in anhydrous ammonia
- High-strength steels (tensile strength of 1240 MPa, or 180 ksi, and above) in a variety of environments, probably with hydrogen embrittlement playing a dominant part
- Copper alloys, notably 70Cu-30Zn cartridge brass by ammonia in humid air

General Features. Stress-corrosion cracks may be intergranular, transgranular, or a combination of both. In aluminum alloys and low-carbon steels, intergranular fracture is usual. High-strength steels and alpha brasses also usually show grain-boundary fracture, with some cracking along matrix twin interfaces in alpha brasses. Transgranular fractures showing extensive branching are typical of SCC in austenitic stainless steels (Fig. 14), and similar transgranular cracks with branches that follow crystallographic planes have been observed in magnesium alloys.

When SCC is transgranular, deviations may occur on a microscopic scale so that the crack may follow microstructural features, such as grain and twin boundaries or specific crystal planes. When SCC is intergranular, the presence of flat, elongated grains means that there is an easy stress-corrosion path normal to the short-transverse direction, which produces "woody" stress-corrosion fracture surfaces. This behavior is typical of extrusions of high-strength aluminum alloys when solution anneal treatment does not cause recrystallization. Some stress-corrosion fractures show progression marks and alternating regions of SCC and overload fracture, with changes in shape of the crack front. The progression marks in the fracture surface shown in Fig. 15 could very easily be confused with fatigue. However, because this high-strength steel part had not been cyclically stressed, it could not be fatigue. The "beach mark" pattern resulted from different rates of penetration of corrosion on the surface as the crack advanced. The final fracture was brittle.

Other features observed on SCC fractures include striations, cleavage facets, and tongues, which can easily be confused with similar features on fatigue and cleavage fractures.

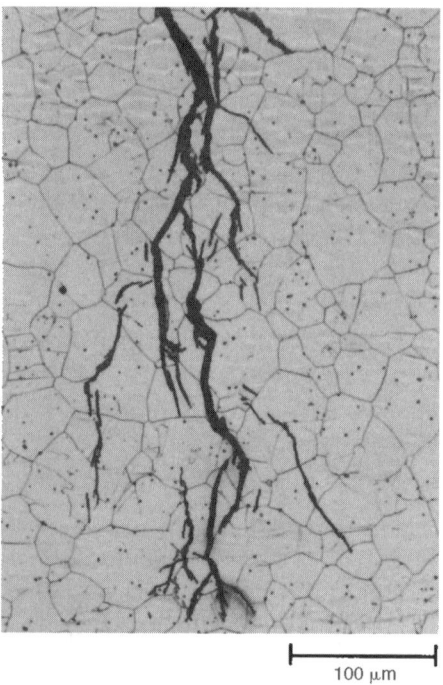

Fig. 14 Micrograph, original magnification at 200×, of a transverse section from a specimen of austenitic stainless steel, showing a branching, transgranular stress-corrosion crack

When intergranular fracture occurs with only superficial corrosion, a rock-candy appearance of the fracture surface is typical. In high-strength aluminum alloys, this kind of fractography defines stress-corrosion fracture. In high-strength steels, this pattern is also characteristic of hydrogen-induced, slow crack growth (see Fig. 16).

Liquid Metal Embrittlement. Metallurgical failure by penetration of liquid metal, usually in grain boundaries, is a unique failure mode that is not often encountered. The penetration of copper alloys by mercury and the penetration of certain steels by molten zinc or copper are typical examples. Such penetration can usually be detected by microscopic examination of polished metallographic sections. Positive identification of the penetrant may be determined by electron-microprobe or SEM-EDX analysis. This type of failure may have characteristic features on the fracture surface in the SEM; that is, it may appear as a brittle intergranular fracture.

Hydrogen-Assisted Cracking. Hydrogen embrittles several metals and alloys, but its deleterious effect on steels, particularly when the strength of the steel is in excess of about 1240 MPa (180 ksi) is most important. A few parts per million of hydrogen dissolved in steel can cause hairline cracking, loss of ductility, and most importantly unexpected reduction to notch toughness. Even when the quantity of gas in solution is too small to reduce ductility, hydrogen-induced delayed fracture (time-delayed cracking) may occur. Gaseous environments containing hydrogen are also damaging. Hairline cracking usually follows prior-austenite grain boundaries and seems to occur when the damaging effect of dissolved hydrogen is superimposed on the stresses that accompany the austenite-to-martensite transformation. Affected areas are recognized on fracture surfaces by their brittle appearance and high reflectivity, which usually contrasts with the

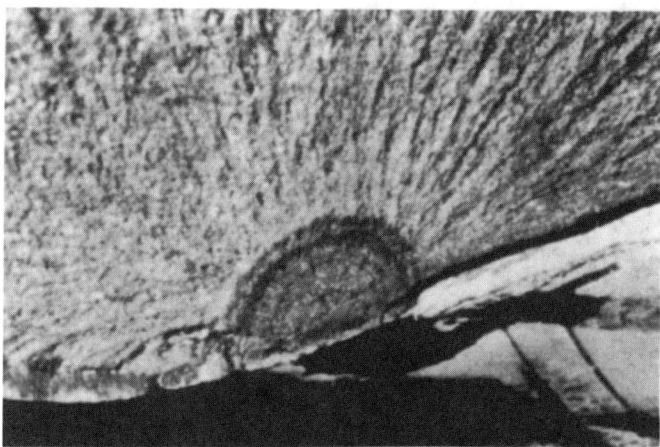

Fig. 15 Fracture surface of a high-strength steel part that failed from stress corrosion, showing progression marks somewhat similar to those observed in fatigue fractures

dull appearance of surrounding regions of ductile fracture. This has led to such areas being described as flakes or "fisheyes."

Hairline cracking is readily recognizable metallographically and is most common near the center of fairly bulky components where constraint of plastic deformation is high, but its incidence may be minimized by modification of steelmaking and heat treating practices or by a change of alloy. Hairline cracking is important with respect to service failures, because such a crack may extend by fatigue and initiate catastrophic fracture.

Hydrogen-embrittled steels, especially those that have suffered delayed fracture, show fracture surfaces very similar to those typical of SCC fracture in aluminum alloys and high-strength steels. In delayed fracture, there is always a region of fracture surface produced by hydrogen-assisted slow crack growth; this crack growth typically follows prior-austenite grain boundaries, as shown in Fig. 8 and 16. Out-of-plane branch cracking along such boundaries is also common. In some steels, hydrogen may promote cleavage fracture. Positive identification is often difficult, and it is frequently impossible to differentiate between hydrogen-induced delayed fracture and SCC cracking fracture.

Creep and Stress Rupture. Creep (time-dependent deformation) is the gradual change in dimension of a metal or alloy under an applied stress at a temperature as low as $0.3\ T_m$, where T_m is the melting point measured on the absolute scale. Thus, lead, tin, and superpure aluminum may de-

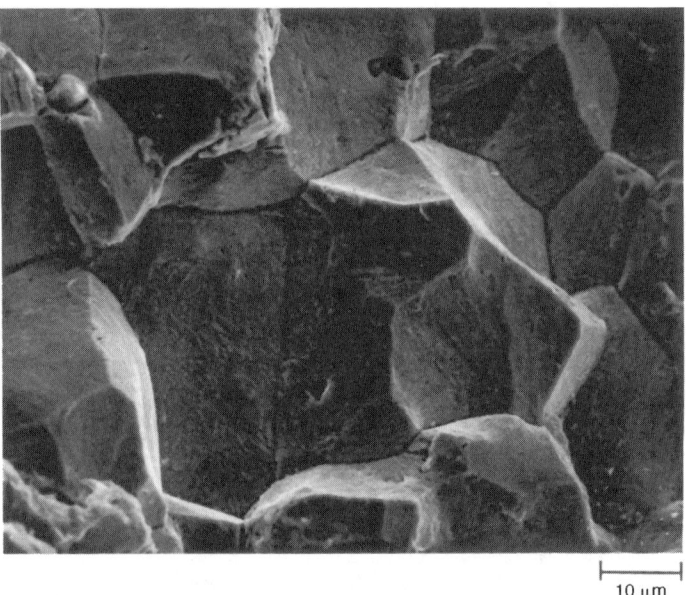

Fig. 16 Example of hydrogen-embrittled steel. Intergranular fracture in an AISI 4130 steel heat treated to an ultimate tensile strength of 1281 MPa (186 ksi) and stressed at 980 MPa (142 ksi) while being charged with hydrogen

form by creep at room temperature or a little above, whereas temperatures near 1000 °C (1832 °F) may be necessary to permit creep in refractory bcc metals such as tungsten and molybdenum and in nickel-base, heat-resisting alloys. Creep strain may produce sufficiently large changes in the dimensions of a component to render it useless for further service before fracture occurs. In other situations, the creep strain may lead to fracture; this type of failure is called stress rupture.

Creep and stress-rupture failures generally are easy to identify; often, they can be recognized by the local ductility and multiplicity of intergranular cracks usually present. Figure 17 shows local deformation and extensive surface cracking associated with a creep failure of a turbine blade. Some superalloys are not very ductile even in a creep failure. For these alloys, stress-rupture failure often can be identified by optical examination of the microstructure, because there is generally a multiplicity of creep voids, usually in the grain boundaries adjacent to the main fracture (see Fig. 18).

Creep and stress-rupture failures are best understood by considering the two general types of creep processes that occur. In the first type, grain-boundary sliding is thought to generate a stress concentration at a triple point that cannot be relieved by plastic deformation in an adjacent grain. This produces a wedge-shaped, grain-boundary crack. The second type involves the initiation of voids at grain boundaries, especially those grain boundaries oriented transversely to a tensile stress, and the growth of the voids by the migration and precipitation of vacancies. This process is called cavitation creep. Stress-rupture fracture due to cavitation creep produces voids that are detectable by metallography or fractography (Fig. 19). Fractography will show that the cavities produced by the voids are not usually spherical, but have complex crystallographic shapes. Striations and terraced patterns may also be observed.

Creep or stress relaxation can also be considered a failure due to slight deformations and not fracture. Stress relaxation of prestressed bolts may lead to leaks or unusual wear. For example, aluminum sheet metal car exteriors have been found to change shape slightly due high residual stresses that result

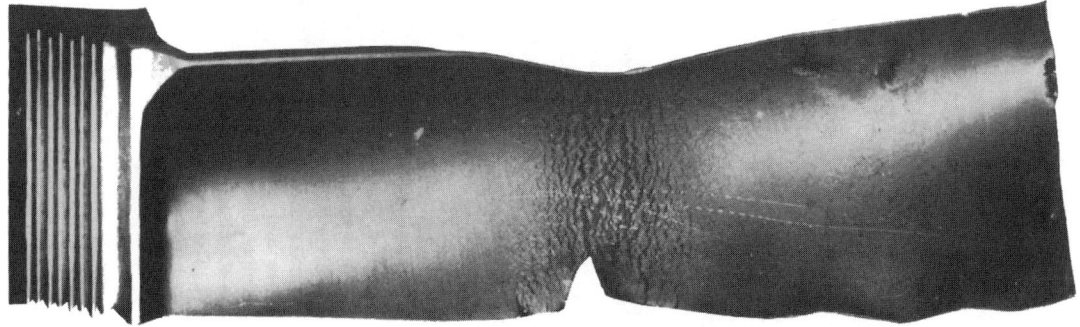

Fig. 17 Typical creep deformation with intergranular cracking in a jet-engine turbine blade

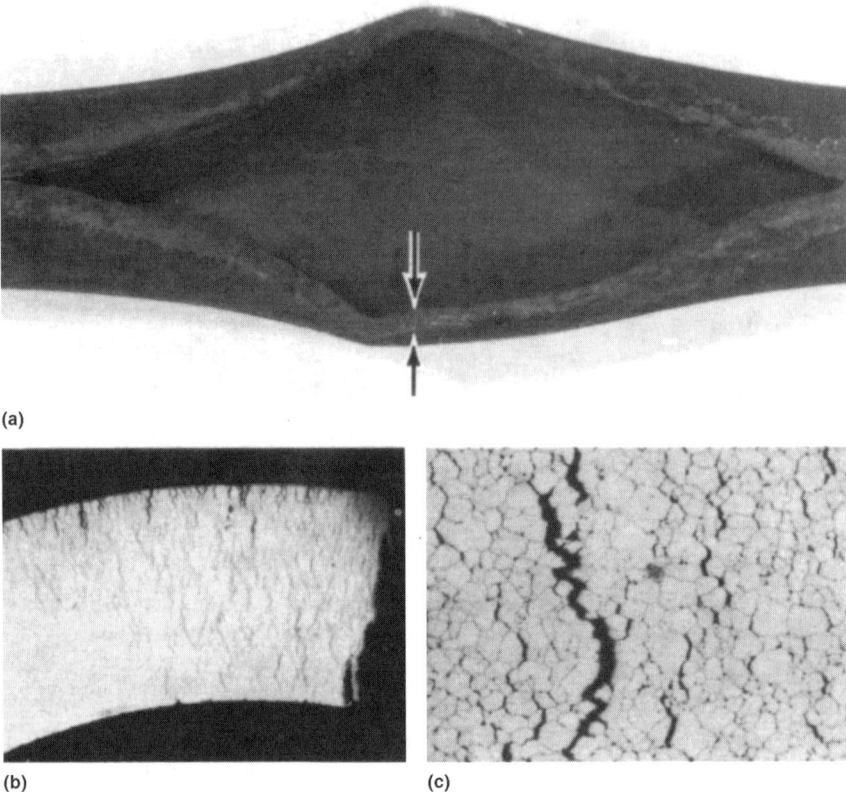

Fig. 18 Type 321 stainless-steel (ASME SA-213, grade TP321H) superheater tube that failed by thick-lip stress rupture. (a) Overall view of rupture. (b) Macrograph of an unetched section from location at arrows showing extensive transverse cracking adjacent to the main fracture (at right). Shown at ~4.5× original magnification. (c) Micrograph of a specimen etched electrolytically in 60% HNO_3 showing intergranular nature of cracking. Original magnification at 100×

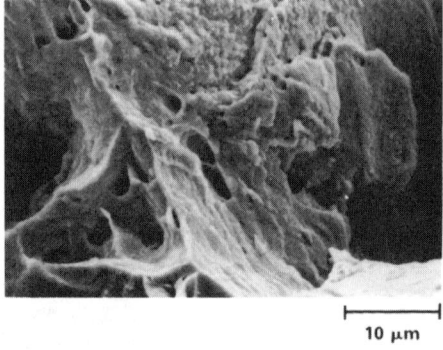

Fig. 19 SEM fractograph of type 316 stainless steel tested in creep to fracture in air at 800 °C (1470 °F) at a load of 103 MPa (15 ksi). Time to rupture, 808 h. The fractograph illustrates the formation of cavities at the grain boundaries. Original magnification at 1260×

in creep at room temperature. This type of failure may not be catastrophic in the sense of a dangerous consequence, but an unwanted wrinkle in the hood of a car is a failure from the car owner's point of view.

Complex Failures. Service failure frequently occurs by the sequential operation of two quite different fracture mechanisms. When conducting a failure analysis, this possibility should always be considered. An example of two types of fracture mechanisms occurring together is shown in Fig. 20. The fractures originated on the inside surface of a drilled hole in an aluminum alloy lug at the points indicated by arrows A and B.

Fractographic examination revealed that the initial cracking was due to stress-corrosion (A and B in Fig. 20), but that crack extension by this mechanism stopped, probably because of internal stress relief. Crack propagation continued by fatigue, as evidenced by beach marks in a band on one of the fracture surfaces (indicated by arrow C in Fig. 20), until catastrophic fracture of the remaining section occurred.

More frequently, a crack may originate due to the fatigue mechanism, then propagate in fatigue until it reaches a critical size, at which time final fracture occurs. As a fatigue crack grows, or progresses, progressive weakening of the part also results, until the remaining thickness cannot support the applied load and the final failure is an overload failure. For metal, overload fracture will have a dimpled fracture surface, but there will be little gross deformation because the fatigue crack acts to limit deformation to the crack tip. Thus, on a macro scale, this type of failure will appear to be brittle.

11. Chemical Analysis

In a failure investigation, routine analysis of the material is usually recommended. Often it is done last because an analysis usually involves destroying a certain amount of material. There are instances where the

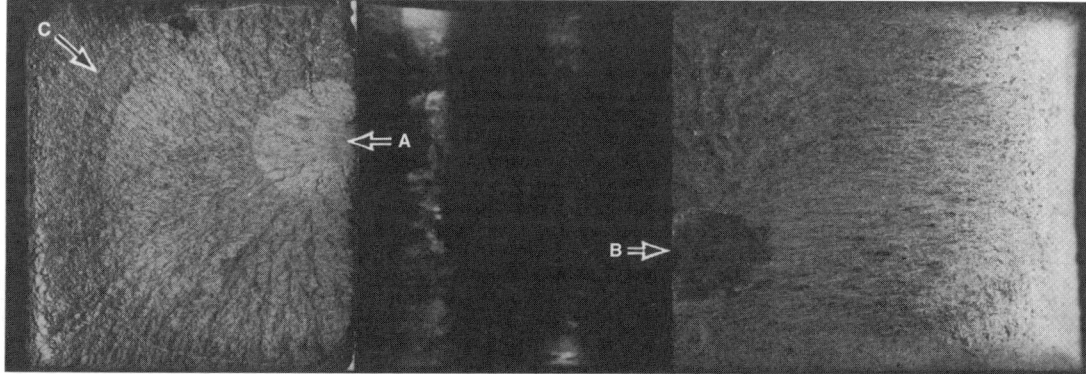

Fig. 20 Fracture surfaces of an aluminum alloy lug. Fractures originated by SCC on the surface of a diametrical hole, at A and B. The crack was then propagated by fatigue, as evidenced by the presence of beach marks at C

wrong material was used, under which conditions the material might be the major cause of failure. In many cases, however, the difficulties are caused by factors other than material composition.

In most instances, slight deviations from specified compositions are not likely to be of major importance in failure analysis. However, small deviations in aluminum content can lead to strain aging in steel, and small quantities of impurities can lead to temper embrittlement. In specific investigations, particularly where corrosion and stress corrosion are involved, chemical analysis of any deposit, scale, or corrosion product, or the substance with which the affected material has been in contact, is required to assist in establishing the primary cause of failure.

Where analysis shows that the content of a particular element is slightly greater than that required in the specifications, it should not be inferred that such deviation is responsible for the failure. Often, it is doubtful whether such a deviation has played even a contributory part in the failure. For example, sulfur and phosphorus in structural steels are limited to 0.04% in many specifications, but rarely can a failure in service be attributed to sulfur content slightly in excess of 0.04%. Within limits, the distribution of the microstructural constituents in a material is of more importance than their exact proportions. An analysis (except a spectrographic analysis restricted to a limited region of the surface) is usually made on drillings representing a considerable volume of material and therefore provides no indication of possible local deviation due to segregation and similar effects.

Also, certain gaseous elements, or interstitials, normally not reported in a chemical analysis, have profound effects on the mechanical properties of metals. In steel, for example, the effects of oxygen, nitrogen, and hydrogen are of major importance. Oxygen and nitrogen may give rise to strain aging and quench aging. Hydrogen may induce brittleness, particularly when absorbed during welding, cathodic cleaning, electroplating, or pickling. Hydrogen is also responsible for the characteristic halos or fisheyes on the fracture surfaces of welds in steels, in which instance the presence of hydrogen often is due to the use of damp electrodes. These halos are indications of local rupture that has taken place under the bursting microstresses induced by the molecular hydrogen, which diffuses through the metal in the atomic state and collects under pressure in pores and other discontinuities. Various effects due to gas absorption are found in other metals and alloys. For example, excessive levels of nitrogen in superalloys can lead to brittle nitride phases that cause failures of highly stressed parts.

Analytical Techniques

Various analytical techniques can be used to determine elemental concentrations and to identify compounds in alloys, bulky deposits, and samples of environmental fluids, lubricants, and suspensions.

Semiquantitative emission spectrography, spectrophotometry, and atomic-absorption spectroscopy can be used to determine dissolved metals (as in analysis of an alloy) with wet chemical methods used where greater accuracy is needed in determining the concentration of metals. Combustion methods ordinarily are used for determining the concentration of carbon, sulfur, nitrogen, hydrogen, and oxygen.

Wet chemical analysis methods are employed for determining the presence and concentration of anions such as Cl^-, NO_3^-, and S^{2-}. These methods are very sensitive.

X-ray diffraction identifies crystalline compounds either on the metal surface or as a mass of particles and can be used to analyze corrosion products and other surface deposits. Minor and trace elements capable of being dissolved can be determined by atomic-absorption spectroscopy of the solution.

The x-ray fluorescence spectrographic technique can be used to analyze both crystalline and amorphous solids, as well as liquids and gases.

Infrared and ultraviolet spectroscopy are used in analyzing organic materials. When the organic materials are present in a complex mixture (such as, for example, solvents, oils, greases, rubber, and plastics), the mixture is first separated into its components by gas chromatography.

Analysis of Surfaces and Deposits. Energy and wavelength-dispersive x-ray (EDX and WDX) spectrometers are frequently used for providing information regarding the chemical composition of surface constituents. They are employed as accessories for SEMs and permit simultaneous viewing and chemical analysis of a surface. If it is desirable to detect the elements in extremely thin surface layers, the Auger electron spectrometer is useful. The Auger electron spectrometer can provide semiquantitative determinations of elements with atomic numbers down to three (lithium). The size of the area examined varies greatly with the test conditions; it may be from 1 to 50 μm in diameter.

For chemical analysis of surface areas as small as 1 μm in diameter, the electron-microprobe analyzer is widely used. This instrument can determine the concentration of all but the low atomic number elements, with a limit of detection below 0.1%. The area examined with the secondary ion mass spectrometer (SIMS technique) is a few microns in diameter larger than that examined with the electron-microprobe analyzer. The ion-microprobe analyzer has the advantage of being able to detect nearly all elements (including those of low atomic weights) in concentrations as low as 100 ppm. It is sometimes used to volatilize materials, which are then passed through a mass spectrometer.

Electron-microprobes and other modern analytical instruments are described in greater detail on pages 32 to 46 in *Failure Analysis and Prevention,* Vol 11 (2002) of the *ASM Handbook.*

The instruments discussed above are used for direct analysis of surfaces; other techniques can be used for analyzing material that has been removed from the surface. For example, if material is removed in a replica (perhaps chemically extracted), it can be analyzed structurally by x-ray diffraction or electron diffraction. Also, depending on the quantity of material extracted, many of the routine chemical analysis techniques may be applicable.

Spot testing uses chemical tests to identify the metal, the alloying elements present, deposits, corrosion products, and soil. Spot tests can be performed both in the laboratory and in the field; they do not require extensive training in analytical chemistry. The only requirement is that the substance be dissolvable; hydrochloric acid or even aqua regia may be used to dissolve the material.

Spot tests for metallic elements such as chromium, nickel, cobalt, iron, and molybdenum are usually done by dissolving a small amount of the alloy in acid and mixing a drop of the resulting solution with a drop of a specific reagent on absorbent paper or a porcelain plate. Spot colorings produced in this way indicate the presence or absence of the metallic radical under test. Samples may be removed from gross surfaces by spotting the specimen with a suitable acid, allowing time for solution and collecting the acid spot with an eye dropper.

For details of methods for detecting and identifying both metallic and nonmetallic elements refer to *Failure Analysis and Prevention,* Volume 11 of the *ASM Handbook* and ASTM STP 550, *Nondestructive Rapid Identification of Metals by Spot Test,* by M.L. Wilson, which gives procedures for spot testing nearly all engineering metals and alloys with a minimum of equipment.

12. Fracture Mechanics Applied to Failure Analysis

The mechanics of fracture in metal parts and specimens under load and the application of fracture mechanics concepts to the design and prediction of service life of parts and components are often pertinent to the investigation of failures due to fracture as well as to the formulation of preventive measures. The concepts of fracture mechanics are useful in measuring fracture toughness and other toughness parameters and in providing a quantitative framework for evaluating structural reliability.

No attempt is made in this appendix to deal with the fundamentals of fracture mechanics. For this, see an introductory discussion in *Principles of Failure Analysis* (Education Course), Lesson 3, or to a detailed discussion in the article "Failure Analysis and Fracture Mechanics," *Failure Analysis and Prevention,* Volume 11 of the *ASM Handbook.* This treatise also provides detailed consideration of notch effects and of toughness testing and evaluation. Among the subjects dealt with in connection with notch effects are stress concentration, triaxiality, plastic constraint, and local strain rate. Toughness testing and evaluation include the fracture toughness test, the dynamic tear test, the crack-opening displacement test, and instrumented impact testing.

13. Simulated-Service Testing

During the concluding stages of an investigation, it may be necessary to conduct tests that simulate the conditions under which failure is believed to have occurred. Often, simulated-service testing is not practical because elaborate equipment is required and even where practical, it is possible that not all of the service conditions are fully known or understood. Corrosion failures, for example, are difficult to reproduce in a laboratory, and some attempts to reproduce them have given misleading results. Serious errors can arise when attempts are made to reduce the time required for a test by artificially increasing the severity of one of the factors—such as the corrosive medium or the operating temperature. Similar problems are encountered in wear testing.

On the other hand, when its limitations are clearly understood, the simulated testing and statistical experimental design analysis of the effects of certain selected variables encountered in service may be helpful in planning corrective action or, at least, may extend service life. The evaluation of the efficacy of special additives to lubricants is an example of the successful application of simulated-service testing. The aircraft industry has made successful use of devices such as the wind tunnel to simulate some of the conditions encountered in flight, and naval architects have employed tank tests to evaluate hull modifications, power requirements, steerage, and other variables that might forestall component failure or promote safety at sea.

Taken singly, most of the metallurgical phenomena involved in failures can be satisfactorily reproduced on a laboratory scale, and the information derived from such experiments can be helpful to the investigator, provided the limitations of the tests are fully recognized. However, many company managers prefer to conduct trials to verify improvements before major conversions are approved. This is a more conservative approach, but it only takes one improper recommendation that results in an adverse result to justify trials of major changes.

14. Consult Other Disciplines

When available, take advantage of the resources of other knowledge and experience available to you. This can be someone in your company or even a vendor. An interdisciplinary approach to complex failure analyses of large structures or machines is often warranted. Activities such as nondestructive testing and FEA are usually subject areas in which the failure analyst is not proficient. Working with engineers and specialists in other disciplines can be required to reach the root cause. Most specialty support such as NDT or FEA can be purchased as a service. However, simple collection of results without a meaningful dialogue of the engineering and metallurgical variables at play in a particular failure may lead to poor results and improper interpretation. Developing a team to solve complex problems can ensure the best possible outcome when the failure requires a broad analysis.

15. *Formulating Conclusions and Report Writing*

At a certain stage in every investigation, the evidence revealed by the examinations and tests outlined in this appendix is analyzed and collated, and premises are formulated. Obviously, many investigations will not become a series of clear-cut stages. If the probable cause of failure is apparent early in the examination, the pattern and extent of subsequent investigation will be directed toward confirmation of the probable cause and the elimination of other possibilities. Other investigations will follow a logical series of stages, as outlined in this appendix, and the findings at each stage will determine the manner in which the investigation proceeds. As new facts modify first impressions, different hypotheses of failure will develop and will be retained or abandoned as dictated by the findings. Where extensive laboratory facilities are available to the investigator, maximum effort will be devoted to amassing the results of mechanical tests, chemical analysis, fractography, and microscopy before the formulation of preliminary conclusions is attempted. Finally, in those investigations in which the cause of failure is particularly elusive, a search through reports of similar instances may be required to suggest possible clues.

Some of the work performed during the course of an investigation may be thought to be unnecessary. It is important, however, to distinguish between work that is unnecessary and that which does not bear fruitful results. During an examination, it is to be expected that some of the work done will not assist directly in determining the cause of failure; nevertheless, negative evidence may be helpful in excluding some potential causes of failure from consideration. Sometimes the evidence that excludes possible failure causes may be stronger than the available evidence supporting the probable cause for a failure.

On the other hand, any tendency to curtail work essential to an investigation should be guarded against. In some instances, it is possible to form an opinion regarding the cause of failure from a single aspect of the investigation, such as visual examination of a fracture surface or examination of a single metallographic specimen. However, before final conclusions are reached, supplementary data confirming the original opinion, if available, should be sought. Total dependence on the conclusions that can be drawn from a single specimen, such as a metallographic section, may be readily challenged unless a history of similar failures can be drawn upon. Yet, circumstances sometimes dictate that a conclusion be formulated on limited data. This is especially true in forensic investigations because of the usually limited extent of the problem and the need to preserve as much evidence as possible. Also, of course, economic limitations are always present. It is rarely justified for the failure analyst to conduct every type of test and investigation possible.

The following checklist, which is in the form of a series of questions, has been proposed as an aid in analyzing the evidence derived from examinations and tests and in formulating conclusions. The questions are also helpful in calling attention to details of the overall investigation that may have been overlooked.

1. Has failure sequence been established?
2. If the failure involved cracking or fracture, have the initiation sites been determined?
3. Did cracks initiate at the surface or below the surface?
4. Was cracking associated with a stress concentration?
5. How long was the crack present?
6. What was the level or magnitude of the loads?
7. What was the type of loading: static, cyclic, or intermittent?
8. Is the fracture surface consistent with the type of loading assumed in the design?
9. What was the failure mechanism?
10. What was the approximate service temperature at the time of failure?
11. Did high or low temperature contribute to failure? Were there temperature excursions?
12. Did wear contribute to failure?
13. Did corrosion contribute to failure? What was the type of corrosion?
14. Was the proper material used? Is a better material required?
15. Was the cross section adequate for the type of service? Were the stresses too high?
16. Was the quality of the material acceptable in accordance with specifications?
17. Were the mechanical properties of the material acceptable in accordance with specifications?
18. Was the component that failed properly heat treated?
19. Was the component that failed properly fabricated?
20. Was the component properly assembled or installed?
21. Was the component repaired during service and, if so, was the repair correctly performed?
22. Was the component properly "run in"?
23. Was the component properly maintained and properly lubricated?
24. Was failure related to abuse in service?
25. Can the design of the component be improved to prevent similar failures?
26. Was the failure primary, or was it damaged by failure of another part?
27. Are failures likely to occur in similar components now in service, and what can be done to prevent their failure?

In general, the answers to these questions will be derived from a combination of records and the examinations and tests previously outlined in this appendix. However, the cause or causes of failure cannot always be determined with certainty. In this instance, the investigation should determine the most probable cause or causes of failure, distinguishing findings based on demonstrated facts from conclusions based on conjecture.

In most cases, the conclusion should be on the first page, closely followed by recommendations.

The failure analysis report should be written clearly, concisely, and logically. One experienced investigator has proposed that the report be divided into the following principal sections:

1. Description of the failed component
2. Service conditions at the time of failure
3. Prior service history
4. Manufacturing and processing history of the component
5. Mechanical structural analysis of the failed part
6. Metallurgical study of the failure
7. Metallurgical evaluation of quality
8. Summary of mechanisms that caused failure
9. Recommendations for prevention of similar failures or for correction of similar components in service
10. Appendix—with figures, tables, and so forth

Obviously, not every report will require coverage under each of these sections. Lengthy reports should begin with an abstract or executive summary. Because readers of failure analysis reports are often people in purchasing, operations, accounting, management, and even legal staff, the avoidance of technical jargon wherever possible is highly desirable. A glossary of terms may also be helpful. The use of appendices, containing detailed calculations, equations, and tables of chemical and metallurgical data, can serve to keep the body of the report clear and uncluttered. The use of literature or book references is recommended along with references to pertinent standards such as ASTM, and so forth.

16. *Follow-Up Recommendations*

The purpose of most industrial failure analysis is to determine the basic or root cause of failure of metal parts. Failures are, at best, a source of irritation and, at worst, a safety hazard, but they are always an economic loss. Therefore, study of the failure should result in carefully formulated recommendations in the report that are aimed at reducing or eliminating similar failures in the future. Such recommendations may involve adding warning labels, changes in design, metallurgy, manufacture, quality control, maintenance, repair practices, and anticipated usage of the product.

If recommendations are made in the report, they should be reviewed with appropriate personnel after the report has been issued so that the recommendations are not overlooked. There may be valid reasons for not implementing the recommendations, but at least those performing the failure analysis and writing the report will have fulfilled their responsibility.

On-Site Failure Analysis

While it is recommended that all stages of a failure analysis be done in a laboratory, there are situations where it is essential to perform failure analyses on the site. To meet the requirements for on-site testing, portable laboratories are available (see Fig. 21). Intended primarily for use in remote locations, the equipment operates either on direct current or on 50 or 60 Hz, 110 V alternating current and requires only a bucket of water and a reasonably clean place in which to work.

Capabilities of the metallography unit include grinding, mechanical polishing, etching, and examination and photographing of macrostructure and microstructure and other features on selected areas of large parts or assemblies. In addition, small specimens can be cut from a part on the site for preparation, examination, and photography immediately or upon return to a fully equipped laboratory. Either mounted or unmounted specimens can be used.

It is frequently desirable to make acetate tape replicas or room-temperature-vulcanizing (RTV) rubber replicas of fracture surfaces or of wear patterns of large parts during an on-site failure analysis. As described in this appendix, several replicas should be made of the fracture origin region using acetate tape softened in acetone, dried, then carefully stripped from the fracture surface. Upon return to the laboratory, the replicas may be

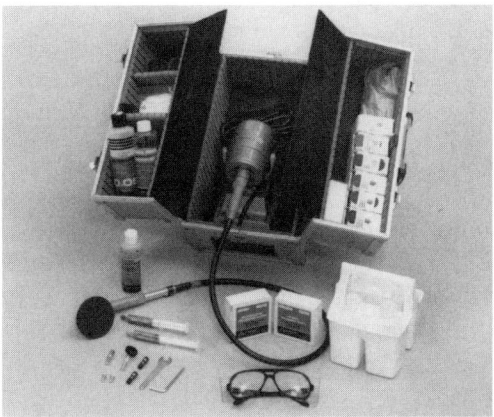

Fig. 21 Arrangement of major components of portable metallographic laboratory for grinding and polishing. Equipment for examination and photography are also available. Image courtesy of Buehler Ltd.

gold-coated and examined in the SEM. Foreign particles removed from the fracture surface also may be analyzed.

The RTV rubber replica can be applied over a rather large area with less chance of missing a critical spot. Combination of acetate tape and RTV rubber replicas can ensure better coverage of the area in question. RTV rubber does not provide the sensitivity of an acetate replica and a setup time of several hours is required. However, the added area can be very important in an investigation.

Hardness testing with a portable hardness testing instrument also may be performed during on-site failure analysis. Several different types of testers are available and in general are either electronic or mechanical in principle. Obviously, small size and light weight are advantages in portable testers.

Equipment and Operation. The major components of the portable laboratory include the following:

- A custom-made machine, plus auxiliary materials, for grinding and polishing small, mounted or unmounted metal specimens
- A right-angle head, electric drill motor with attachments and materials for grinding and polishing selected spots on large parts or assemblies; the drill motor also is used for driving the grinding and polishing machine described in the first item listed above
- A portable microscope, with camera attachment and film for use in photographing metallographic specimens
- Equipment and materials for mounting and etching specimens
- A handheld single-lens reflex 35 mm camera, with macrolenses and film
- A pocket-size magnifier, and a ruler or scale
- A hacksaw and blades for cutting specimens
- Portable hardness tester
- Acetate tape, acetone, and containers
- RTV rubber for replicas

Moral, Ethical, and Legal Responsibility

In most cases the failure analyst has a clear interest in reducing or eliminating the failure cause. However, in all forensic investigations there are two sides. One investigator may be defending a client's or their own company's position. The analyst must always be truthful and base his or her opinion on fact and sound engineering judgment. Above all, the analyst must point out clearly dangerous or hazardous conditions even if it may seem to hurt the analyst's employer or client. In the final analysis, identifying and correcting a problem as soon as possible is the correct business and ethical choice.

Summary

1. There are 16 possible steps to follow in conducting a failure analysis-from collecting and examining samples to testing, determining the failure mode, writing the report, and developing follow-up recommendations.
2. In working to produce economical, reliable products, it is vital for manufacturers to recognize the importance of analyzing failures that occur during product development testing or in service.
3. Of primary importance in conducting a thorough failure analysis, is the compilation of information regarding the manufacturing, processing, and service history of the failed part. This includes obtaining original specifications and drawings, if available.
4. It is important for failure analysts to keep written and visual records of their investigations.
5. An essential first step in any failure analysis is a thorough visual examination including consideration of fluids, soil, debris, paint, or corrosion found on the surface of the failed part.
6. Nondestructive testing methods—magnetic-particle, liquid-penetrant, ultrasonic, radiography, and eddy current testing—as well as mechanical tests—hardness, tensile, and impact—can be helpful in determining the cause of a failure.
7. The influence of the chemical and thermal environment must be kept in mind during any failure analysis. Corrosion and property changes may be critical in an analysis.
8. Macroscopic examination of the fracture surfaces can be considered the most important part of a failure analysis. It is typically performed at magnifications of 1 to $100\times$ using the unaided eye, a hand lens, a low-power stereoscopic microscope, or SEM.
9. Performing both macro- and microscale microstructural examination of unetched and etched specimens provides important evidence about the type and structure of the material involved in the failure.
10. Brittle fractures are typically more difficult to analyze because of the large number of possible fracture mechanisms and the lack of obvious deformation.
11. In any failure analysis, all avenues must be explored to ensure that the investigator's analysis is complete and that the final conclusions are well documented and accurate.

SELECTED REFERENCES FOR FURTHER STUDY

- J.A. Collins, Modes of Mechanical Failure, *Failure of Materials in Mechanical Design,* John Wiley & Sons, 1981, p 6–14
- J.A. Collins, The Role of Failure Prevention Analysis in Mechanical Design, *Failure of Materials in Mechanical Design,* John Wiley & Sons, 1981, p 1–5

- *Failure Analysis and Prevention,* Vol 11, *ASM Handbook,* ASM International, 2002
- *Fractography,* Vol 12, *ASM Handbook,* ASM International, 1987
- G.F. Vander Voort, Conducting the Failure Examination, *Met. Eng. Quart.,* Vol 15, 1975
- G.F. Vander Voort, Macroscopic Examination Procedures for Failure Analysis, *Metallography in Failure Analysis,* J.L. McCall and P.M. French, Ed., Plenum Press, 1978
- D.J. Wulpi, Techniques of Failure Analysis, *Understanding How Components Fail,* 2nd ed., ASM International, 1999, p 1–11

APPENDIX 2

Glossary

A

abrasion. The process of grinding or wearing away through the use of abrasives; a roughening or scratching of a surface due to *abrasive wear*.

abrasive wear. The removal of material from a surface when hard particles slide or roll across the surface under pressure. The particles may be loose or may be part of another surface in contact with the surface being abraded. Compare with *adhesive wear*.

addendum. That portion of a gear tooth between the pitch line and the tip of the tooth. Plural is "addenda."

adhesive wear. The removal or displacement of material from a surface by the welding together and subsequent shearing of minute areas of two surfaces that slide across each other under pressure. Compare with *abrasive wear*.

age hardening. Hardening by aging, usually after rapid cooling or cold working. See *aging*.

aging. A change in the properties of certain metals and alloys that occurs at ambient or moderately elevated temperatures after hot working or a heat treatment (quench aging in ferrous alloys, natural or artificial aging in ferrous and nonferrous alloys) or after a cold working operation (strain aging). The change in properties is often, but not always, due to a phase change (precipitation), but never involves a change in chemical composition of the metal or alloy.

ambient. Surrounding. Usually used in relation to temperature, as "ambient temperature" surrounding a certain part or assembly.

annealing. A generic term denoting a treatment—heating to and holding at a suitable temperature followed by cooling at a suitable rate—used primarily to soften metallic materials, but also to produce desired changes simultaneously in other properties or in microstructure. When applied only for the relief of stress, the process is called stress relieving or stress-relief annealing.

anode. The electrode at which oxidation or corrosion occurs. It is the opposite of *cathode*.

anodizing. Forming a surface coating for wear protection or aesthetic purposes on a metal surface. Usually applied to aluminum, in which an aluminum oxide coating is formed in an electrolytic bath.

applied stress. The stress applied to a part or assembly as a result of external forces or loads.

arc strike. The location where a welding electrode has contacted a metal surface, melting a small volume of metal.

asperity. A peak or projection from one surface. Used as a term in wear technology or tribology.

austenite. An elevated-temperature parent phase in ferrous metals from which all other low-temperature structures are derived. The normal condition of certain types of stainless steels.

axial. Longitudinal, or parallel to the axis or centerline of a part. Usually refers to axial compression or axial tension.

B

bainite. An intermediate transformation product from *austenite* in the heat treatment of steel. Bainite can somewhat resemble *pearlite* or *martensite,* depending on the transformation temperature.

beachmarks. Macroscopic (visible) lines on a fatigue fracture that show the location of the tip of the fatigue crack at some point in time. Must not be confused with *striations,* which are extremely small and are formed in a different way.

biological corrosion. Deterioration of metals as a result of the metabolic activity of microorganisms. Also known as microbiological corrosion.

blistering. See *hydrogen blistering*.

body-centered cubic. See *cell*.

brittle. Permitting little or no plastic (permanent) deformation prior to fracture.

brittleness. The tendency of a material to fracture without first undergoing *plastic deformation*. Contrast with *ductility*.

C

carbonitriding. An elevated-temperature process (similar to *carburizing*) by which a ferrous metal absorbs both carbon and nitrogen into the surface when exposed to an atmosphere high in carbon and nitrogen. The carbon and nitrogen atoms actually diffuse, or flow, into the metal to form a high-carbon, high-nitrogen zone near the surface.

carburizing. An elevated-temperature process by which a ferrous metal absorbs carbon into the surface when exposed to a high-carbon environment. Carbon atoms actually diffuse, or flow, into the metal to form a high-carbon zone near the surface.

case. In a ferrous metal, the outer portion that has been made harder than the interior, or *core*. The case is usually formed by diffusion of other atoms—particularly carbon and/or nitrogen—into the metal, but may also be formed by localized heat treating of the surface, as by flame or induction hardening.

case crushing. See *subcase fatigue*.

case depth. The depth of the case, or hardened surface region, of a metal, usually steel. Since there are many ways of determining case depth, the method used should be stated.

cathode. The electrode at which reduction (and practically no corrosion) occurs. It is the opposite of *anode*.

cathodic protection. Reduction or elimination of corrosion by making the metal a cathode by means of an impressed direct current or attachment of a sacrificial anode (usually magnesium, aluminum, or zinc).

cause-and-effect diagram. See *fishbone diagram*.

caustic embrittlement. Cracking as a result of the combined action of tensile stresses (applied or residual) and corrosion in alkaline solutions (as at riveted joints in boilers).

cavitation pitting fatigue. A type of pitting fatigue in which cavities, or regions of negative pressure, in a liquid implode, or collapse inward, against a metal surface to cause pits or cavities in the metal surface if repeated often enough at the same points on the surface.

cell. (1) A "building block" forming a grain or crystal. The cell (or "unit cell") is composed of a small number of atoms arranged in any of several different configurations, depending on the metal. The most common are cubic (with an atom at each corner); body-centered cubic (same as cubic, but also has an atom at the center of the cube); face-centered cubic (same as cubic, but also has an atom at the center of each face, or side); hexagonal; and tetragonal. (2) An electrical circuit consisting of an anode and a cathode in electrical contact with a solid or liquid electrolyte. Corrosion generally occurs only at anodic areas.

Charpy test. An impact test in which a V-notched, keyhole-notched, or U-notched specimen, supported at both ends, is struck behind the notch by a striker mounted at the lower end of a bar that can swing as a pendulum. The energy that is absorbed in fracture is calculated from the height to which the striker would have risen had there been no specimen and the height to which it actually rises after fracture of the specimen.

chromizing. An elevated-temperature process by which a ferrous metal absorbs chromium into the surface when exposed to a high-chromium environment. Chromium atoms actually diffuse, or flow, into the metal to form a high-chromium surface layer.

circumferential. Around the circumference, or periphery, of a circle or a cylinder like a wheel or a shaft. Also called *tangential* or *hoop* when referring to stresses.

clad metal. A composite metal containing two or three layers that have been bonded together. The bonding may have been accomplished by rolling together, welding, casting, heavy chemical deposition, or heavy electroplating.

cleavage fracture. Splitting fracture of a metal along the edges of the cells but across the grains, or crystals. This is a brittle *transgranular fracture,* contrasted to a brittle *intergranular fracture,* in which the fracture is between the grains.

clevis joint. A U-shaped part with holes for a pin to hold another part between the sides of the U.

cohesive strength. The force that holds together the atoms in metal crystals. Analogous to *tensile strength,* but on a submicroscopic scale.

cold heading. Axial compression of the end of a metal cylinder to enlarge the cross section. Used to form the head of a nail or bolt.

cold shut. (1) A discontinuity that appears on the surface of cast metal as a result of two streams of liquid metal meeting but failing to unite. (2) A lap on the surface of a forging or billet that was closed without fusion during deformation. Same as forging lap.

cold work. Permanent deformation caused by application of an external force to a metal below its recrystallization temperature.

compressive. Pertaining to forces on a body or part of a body that tend to crush, or compress, the body.

compressive strength. In compression testing, the ratio of maximum load to the original cross-sectional area. Fracture may or may not occur, depending on the applied forces and the properties of the material.

concentration cell. A cell involving an electrolyte and two identical electrodes, with the electrical potential resulting from differences in the chemistry of the environments adjacent to the two electrodes.

conformal. Describing two surfaces that conform to each other; that is, they nest together, as does a convex surface that fits within a concave surface. Example: a bearing ball within an inner or outer *raceway.* Compare with *counterformal.*

core. In a ferrous metal, the inner portion which is softer than the exterior, or *case.*

corrosion. Deterioration of a metal by chemical or electrochemical reaction with its environment.

corrosion fatigue. The combined action of corrosion and fatigue (cyclic stressing) in causing metal fracture.

counterformal. Describing two convex surfaces that are in contact but do not nest together. Examples: two gear teeth; also, a roller bearing against an inner *raceway.* Compare with *conformal.*

crack growth. Rate of propagation of a crack through a material due to a static or dynamic applied load.

crack opening displacement (COD). On a K_{Ic} specimen, the opening displacement of the notch surfaces at the notch and in the direction perpendicular to the plane of the notch and the crack. The displacement at the tip is called the crack tip opening displacement (CTOD); at the mouth, it is called the crack mouth opening displacement (CMOD).

crack size. A lineal measure of a principle planar dimension of a crack. This measure is commonly used in the calculation of quantities descriptive of the stress and displacement fields. In practice, the value of crack size is obtained from procedures for measurement of physical crack size, original crack size, or effective crack size, as appropriate to the situation under consideration.

crack-tip plane strain. A stress-strain field near a crack tip that approaches *plane strain* to the degree required by an empirical criterion.

creep. Time-dependent strain occurring under stress. Or, change of shape that occurs gradually under a steady load.

creep-rupture strength. The stress that causes fracture in a creep test at a given time, in a specified constant environment. Sometimes referred to as the stress-rupture strength.

crevice corrosion. Localized corrosion resulting from the formation of a *concentration cell* in a crevice between a metal and a nonmetal, or between two metals.

crystal. A three-dimensional array of atoms having a certain regularity in its internal arrangement. The crystal is composed of many cells, or lattices, in which the atoms are arranged in a given pattern, depending on the metal involved. Another name for crystal is *grain,* which is more commonly used in practical metallurgy.

crystallographic. Pertaining to the crystal structure of a metal.

cyclic load. Repetitive loading, as with regularly recurring stresses on a part, that sometimes leads to fatigue fracture.

cyclic stress. Same as *cyclic load.*

D

decarburization. Loss of carbon from the surface of a ferrous (iron-base) alloy as a result of heating in a medium that reacts with the carbon at the surface.

dedendum. That portion of a gear tooth between the pitch line and the root of the tooth. Plural is "dedenda."

deflection. Deformation within the elastic range caused by a load or force that does not exceed the *elastic limit* of the material. Temporary deformation, as that of a spring.

deoxidized metal. Metal that has been treated, when in the liquid state, with certain materials that tend to form oxides, thus removing the oxygen from the metal.

design of experiments (DOE). A methodology involving statistically designed experiments in which the character and sequence of individual experiments are planned in advance so that data are taken in a way that will provide the most unbiased and precise results commensurate with the available time and money.

dimpled rupture fracture. A fractographic term describing ductile fracture that occurs by the formation and coalescence of microvoids along the fracture path. Seen at high magnification as tiny cups, or half-voids.

distortion. Change in the shape of a part due to the action of mechanical forces. Excludes removal of metal by wear or corrosion.

ductile. Permitting plastic (or permanent) deformation prior to eventual fracture.

ductility. The ability of a material to deform plastically before fracturing. Measured by elongation or reduction of area in a tension test, by height of the cup formed in a cupping test, or by the radius or angle of bend in a bend test. Contrast with *brittleness*; see also *plastic deformation*.

dynamic. Moving, or having high velocity. Frequently used with impact testing of metal specimens. Opposite of *static,* or essentially stationary, testing or service.

E

elastic. Able to return immediately to the original size and shape after being stretched or squeezed; springy.

elasticity. The property of being *elastic*.

elastic limit. The maximum stress a material is capable of sustaining without any permanent shape change remaining on complete release of the stress.

elastic-plastic fracture mechanics. A design approach used for materials that fracture or behave in a "plastic" manner, such as lower strength, high-toughness steels.

elastomer. A material with rubberlike properties—that is, quite elastic, returning to its original size and shape after being deformed.

electrochemical. Pertaining to combined electrical and chemical action. Deterioration (corrosion) of a metal occurs when an electrical current flows between cathodic and anodic areas on metal surfaces.

electrode. An electrical conductor, usually of metal or graphite, that leads current into or out of a solution (electrolyte).

electrolyte. A material, usually a liquid or paste, that will conduct an electric current.

embrittlement. The severe loss of *ductility* and/or toughness of a material. See also *hydrogen embrittlement*.

endurance limit. See *fatigue limit*.

ε-N curve. Plot of strain versus number of load cycles indicating fatigue behavior of a metal test specimen, which takes into account both elastic and plastic responses to applied loadings.

erosion. (1) Loss of material from a solid surface due to relative motion in contact with a fluid that contains solid particles. Erosion in which the relative motion of particles is nearly parallel to the solid surface is called abrasive erosion. Erosion in which the relative motion of the solid particles is nearly normal to the solid surface is called impingement erosion or impact erosion. (2) Progressive loss of original material from a solid surface due to mechanical interaction between that surface and a fluid, a multicomponent fluid, and impinging liquid, or solid particles.

erosion-corrosion. A conjoint action involving corrosion and erosion in the presence of a corrosive substance.

eutectic alloy. An alloy having the composition indicated by the relatively low melting temperature on an equilibrium diagram of two metals.

F

face-centered cubic. See *cell*.

failure. Cessation of function or usefulness of a part or assembly. The major types of failure are *corrosion, distortion, fracture,* and *wear*.

failure mode assessment (FMA) chart. A spreadsheet used in a failure investigation to assess the probability of and assign priority to potential root causes of failure. An FMA chart is developed from the *fault tree*. A completed FMA chart is used to develop a *technical plan for resolution (TPR) chart*.

failure modes and effects analysis (FMEA). A methodology for analyzing potential reliability problems early in the development cycle where it is easier to take actions to overcome these issues, thereby enhancing reliability through design. FMEA is used to identify potential failure modes, determine their effect on the operation of the product, and identify actions to mitigate the failures.

false brinelling. Fretting wear between a bearing ball and its *raceway*. Makes a dark depression in the race, similar to that made by an indentation from a Brinell hardness test. Properly called *fretting wear*.

fatigue. The phenomenon leading to fracture under repeated or fluctuating stresses having a maximum value less than the *tensile strength* of the material. Fatigue fractures are progressive, beginning as minute cracks that grow under the action of the fluctuating stresses.

fatigue life. The number of cycles of stress that can be sustained prior to failure for a stated condition.

fatigue limit. The maximum stress below which a material can presumably endure an infinite number of stress cycles. If the stress is not completely reversed, the value of the mean stress, the minimum stress, or the stress ratio should be stated.

fatigue strength. The maximum stress that can be sustained for a specified number of cycles without failure, the stress being completely reversed within each cycle unless otherwise stated.

fault tree. A method that provides a systematic description of the combinations of possible occurrences in a system that can result in failure. It is a graphical representation of the Boolean logic that relates to the output (top) event.

ferrite. Essentially pure iron in the microstructure of an iron or steel specimen. It may have a small amount of carbon (less than 0.02 wt%). Also called alpha iron.

ferrous. Describing a metal that is more than 50% iron, such as steel, stainless steel, cast iron, ductile (nodular) cast iron, etc.

fillet. A radius (curvature) imparted to inside meeting surfaces; a blended curve joining an internal corner to two lateral surfaces.

fishbone diagram. A systematic analysis tool that organizes the effects of a problem and its possible causes, in a graphical display that often resembles the skeleton of a fish. Also known as a cause-and-effect diagram.

forensic. Pertaining to the employment of scientific technology to assist in the determination of facts in courts of law.

fractographic. Pertaining to photographic views of fracture surfaces, usually at high magnification.

fracture. A break, or separation, of a part into two or more pieces.

fracture mechanics. A quantitative analysis for evaluating structural behavior in terms of applied stress, crack length, and specimen or machine component geometry.

fracture toughness. A generic term for measures of resistance to extension of a crack. The term is sometimes restricted to results of *fracture mechanics* tests, which are directly applicable in fracture control. However, the term commonly includes results from simple tests of notched or precracked specimens not based on fracture mechanics analysis. Results from tests of the latter type are often useful for fracture control, based on either service experience or empirical correlations with fracture mechanics tests.

free-body diagram. A rectangle representing a theoretical point on the surface of a part under stress that shows, in a simplified way, the stresses and components of stresses acting on the part.

fretting wear. Surface damage to a metal part resulting from microwelding due to slight movement in a nearly stationary joint. Also called fretting corrosion.

G

galvanic corrosion. Corrosion associated with the current of a galvanic cell consisting of two dissimilar conductors in an electrolyte or two similar conductors in dissimilar electrolytes. Where the two dissimilar metals are in contact, galvanic corrosion may occur.

galvanic series. A series of metals and alloys arranged according to their relative corrosive tendency in a given environment. The most common environment is seawater or other concentrations of salt in water.

gas porosity. A cavity caused by entrapped gas. Essentially a smooth-sided bubble within the metal, where the metal solidified before the gas could escape to the atmosphere. Also called gas pocket.

general corrosion. See *uniform corrosion.*

gradient. A slope, such as a temperature gradient across a part in which one side is hotter than the other.

grain. The more common term for *crystal,* a three-dimensional array of atoms having a certain regularity in its internal arrangement. The grain is composed of many cells, or lattices, in which the atoms are arranged in a given pattern, depending on the metal involved.

grain boundary. The boundary between two grains.

graphitization. Formation of graphite in iron or steel, caused by precipitation of carbon from the iron-carbon alloy.

H

halides. A group of compounds containing one of the halogen elements—bromine, chlorine, fluorine, or iodine—that are sometimes damaging to metals. One of the most common halide compounds is sodium chloride, or ordinary salt.

hardness. Resistance of metal to *plastic deformation,* usually by indentation. However, the term may also refer to stiffness or temper, or to resistance to scratching, abrasion, or cutting. Indentation hardness may be measured by various hardness tests, such as Brinell, Rockwell, Knoop, and Vickers. All indentation hardness tests employ arbitrary loads applied to arbitrarily shaped indentors, or penetrators.

HB. Abbreviation for "Hardness, Brinell," a hardness test. The number relates to the applied load and to the surface area of the permanent impression in a metal surface made by a hardened steel or carbide ball. Also known as "BHN" or Brinell hardness number.

heat treatment. Heating and cooling a metal or alloy in such a way as to obtain desired conditions or properties.

high-cycle fatigue. Fatigue that occurs at relatively large numbers of cycles, or stress applications. The numbers of cycles may be in the hundreds of thousands, millions, or even billions. There is no exact dividing line between low- and high-cycle fatigue, but for practical purposes, high-cycle fatigue is not accompanied by plastic, or permanent, deformation.

Hooke's law. Stress is proportional to strain. This law is valid only up to the proportional limit, or the end of the straight-line portion of the stress-strain curve.

hoop. See *circumferential*.

hot heading. Axial compression of the end of a metal cylinder at an elevated temperature to enlarge the cross section. Also called upsetting.

HRB, HRC. Abbreviations for "Hardness, Rockwell B" and "Hardness, Rockwell C" respectively. The Rockwell B and C scales are two indentation hardness scales commonly used with metals. All Rockwell scales measure the depth of penetration of a diamond or hardened steel ball that is pressed into the surface of a metal under a standardized load.

hydrogen blistering. The formation of blisters on or below a metal surface from excessive internal hydrogen pressure. Hydrogen may be formed during cleaning, plating, corrosion, and so on.

hydrogen embrittlement. A condition of low *ductility* or cracking in metals resulting from the absorption of hydrogen.

hydrostatic. Describing three-dimensional compression similar to that imposed on a metal part immersed in a liquid under pressure.

hypoid. A type of bevel, or conical, gear in which the teeth are extremely curved within the conical shape. The teeth of the pinion, or driving gear, are more curved than a spiral bevel pinion, tending to wrap around the conical shape.

I

impingement. A process resulting in a continuing succession of impacts between liquid or solid particles and a solid surface.

impingement erosion. Loss of material from a solid surface due to *impingement*.

implode. Burst inward, such as in a collapsing cavity, or negative-pressure region, during *cavitation pitting fatigue*.

inclusions. Nonmetallic particles, usually compounds, in a metal matrix. Usually considered undesirable, though in some cases, such as in free machining metals, inclusions may be deliberately introduced to improve machinability.

induction hardening. A method of locally heating the surface of a steel or cast iron part through the use of alternating electric current. It is usually necessary to rapidly cool, or quench, the heated volume to form *martensite,* the desired hard microstructure.

interface. The boundary between two contacting parts or regions of parts.

intergranular corrosion. Corrosion that occurs preferentially at or immediately adjacent to grain boundaries (usually with slight or negligible attack in the grain interior).

intergranular fracture. Brittle fracture of a metal in which the fracture is between the grains, or crystals, that form the metal. Contrast with *transgranular fracture*.

intermetallic phase precipitation. Formation of a very large number of particles of an intermediate phase in an alloy system.

K

keyway. A longitudinal groove, slot, or other cavity usually in a shaft, into which is placed a key to help hold a hub on the shaft. The key and keyway are used for alignment and/or mechanical locking.

L

lamellar. Plate-like; made of a number of parallel plates or sheets. Usually applied to microstructures. The most common lamellar microstructure is *pearlite* in ferrous metals.

lateral. In a sideways direction.

lattice, lattice structure. Same as *cell*.

linear elastic fracture mechanics. A method of fracture analysis that can determine the stress (or load) required to induce fracture instability in a structure containing a crack-like flaw of known size and shape.

liquid metal embrittlement. The decrease in ductility and toughness of a metal caused by contact with another metal in liquid form. Results in intergranular fracture. Also known as liquid metal induced embrittlement. See also *solid metal embrittlement*.

longitudinal. Lengthwise, or in an *axial* direction.

low-cycle fatigue. Fatigue that occurs at relatively small numbers of cycles, or stress applications. The numbers of cycles may be in the tens, hundreds, or even thousands of cycles. There is no exact dividing line between low- and high-cycle fatigue, but for practical purposes, low-cycle fatigue may be accompanied by some plastic, or permanent, deformation.

M

macroscopic. Visible at magnifications up to about 25–50 times.

martensite. The very hard structure in certain irons and steels that is usually formed by quenching (rapid cooling) from an elevated temperature. Martensite may or may not be tempered to reduce hardness and increase ductility and toughness.

martensitic transformation. Formation of *martensite*.

matrix. The principal phase of a metal in which another constituent is embedded. For example, in gray cast iron, the metal is the matrix in which the graphite flakes are embedded.

mechanical properties. The properties of a material that reveal its elastic and inelastic (plastic) behavior when force is applied, thereby indicating its suitability for mechanical (load-bearing) applications. Examples are elongation, *fatigue limit, hardness, modulus of elasticity, tensile strength,* and *yield strength.*

metal. An opaque, lustrous elemental chemical substance that is a good conductor of heat and electricity and, when polished, a good reflector of light. Most elemental metals are malleable and ductile and are, in general, heavier than the other elemental substances.

metallographic. Pertaining to examination of a metallic surface with the aid of a microscope. The surface is usually polished to make it flat, and may be etched with various chemicals to reveal the microstructure.

microbiological corrosion. See *biological corrosion.*

microscopic. Visible only at magnifications greater than about 25–50 times.

microstructure. The structure of polished and etched metals as revealed by a microscope at a magnification greater than 25–50 times.

microvoid. A microscopic cavity that forms during fracture of a ductile metal. A very large number of microvoids form in the region with the highest stress; some of them join together to form the actual fracture surface, each side of which contains cuplike half-voids, usually called dimples.

mode. One of the three classes of crack (surface) displacements adjacent to the crack tip. These displacement modes are associated with stress-strain fields around the crack tip.

modulus of elasticity. A measure of the stiffness of a metal in the elastic range—that is, the degree to which a metal will deflect when a given load is imposed on a given shape. Also called Young's modulus.

monomolecular. Describing a film or surface layer one molecule thick.

monotonic. Pertaining to a single load application in a relatively short time, as in a monotonic tensile test. Same as *static.*

N

necking. The reduction in cross-sectional size that occurs when a part is stretched by a tensile stress.

nitriding. An elevated-temperature process (but lower than carburizing or carbonitriding) by which a ferrous metal absorbs nitrogen atoms into the surface when exposed to a high-nitrogen environment. Nitrogen atoms actually diffuse, or flow, into the metal to form a high-nitrogen surface layer.

nonferrous. Describing a metal that is less than 50% iron, such as aluminum, copper, magnesium, and zinc and their alloys.

normal stress. See *stress.*

notch. See *stress concentration.*

notched-bar impact test. A standardized mechanical test in which a metal test specimen with a specified notch is struck with a standardized swinging pendulum weight. The type of fracture and the energy absorbed by the fracturing process can be determined from the specimen.

notch toughness. An indication of the capacity of a metal to absorb energy when a *notch,* or stress concentrator, is present.

O

overload. A condition in which a system is given too high of an input level. A common cause of distortion or product failure.

P

pancake forging. Plastic deformation of a very ductile material under axial compressive forces between flat, parallel dies. The sides bulge outward, while the other surfaces become essentially flat and parallel.

Pareto diagram. A frequency diagram used to plot the relative importance of the differences between groups of data in the form of a bar chart. Items with the greatest to lowest frequency are plotted from left to right. The diagram visually exposes the relative magnitude of items, providing a base of uniform knowledge from which to solve the problem.

Paris equation. A generalized fatigue-crack-growth rate exponential-power law that shows the dependence of fatigue-crack-growth rate on the stress-intensity factor, K, and has been verified by many investigations.

pearlite. A lamellar, or platelike, microstructure commonly found in steel and cast iron.

physical properties. The properties of a material that are relatively insensitive to structure and can be measured without the application of force. Examples are density, melting temperature, damping capacity, thermal conductivity, thermal expansion, magnetic properties, and electrical properties.

pitch line. The location on a gear tooth, approximately midway up the tooth, that crosses the pitch circle, or the equivalent-size disk that could geometrically replace the gear.

pitting. (1) Corrosion of a metal surface, confined to a point or small area, that takes the form of cavities. (2) In tribology, a type of wear characterized by the presence of surface cavities formed by processes such as fatigue, local adhesion, or cavitation.

plastic deformation. The permanent (inelastic) distortion of metals under applied stresses that strain the material beyond its *elastic limit.*

polycrystalline. Pertaining to a solid metal composed of many crystals, such as an ordinary commercial metal.

polymeric. Pertaining to a polymer, or plastic.

poultice corrosion. Same as *crevice corrosion,* but usually applies to a mass of particles or an absorptive material in contact with a metal surface that is wetted periodically or continuously. Corrosion occurs under the edges of the mass of particles or the absorptive material that retains moisture.

precipitation hardening. Hardening caused by the precipitation of a constituent from a supersaturated solid solution. See also *age hardening* and *aging.*

prestress. Stress on a part or assembly before any service or operating stress is imposed. Similar to internal or *residual stress.*

primary creep. The first, or initial, stage of *creep,* or time-dependent deformation.

proportional limit. The maximum stress at which strain remains directly proportional to stress; the upper end of the straight-line portion of the stress-strain or load-elongation curve.

psi. Abbreviation for pounds per square inch, a unit of measurement for stress, strength, and modulus of elasticity.

Q

quasi-cleavage fracture. A fracture mode that combines the characteristics of cleavage fracture and dimpled rupture fracture. An intermediate type of fracture found in certain high-strength metals.

quenching. Rapid cooling of metals (often steels) from a suitable elevated temperature. This generally is accomplished by immersion in water, oil, polymer solution, or salt, although forced air is sometimes used.

R

raceway. The tracks or channels on which roll the balls or rollers in an antifriction rolling-element bearing. The inner race fits around a shaft, while the outer race fits within a hole in a larger part.

radial. In the direction of a radius between the center and the surface of a circle, cylinder, or sphere.

ratchet marks. Ridges on a fatigue fracture that indicate where two adjacent fatigue areas have grown together. Ratchet marks usually originate perpendicular to a surface and may be straight or curved, depending on the combination of stresses that is present.

reactive metals. Metals that tend to react with the environment, usually those near the anodic end of the galvanic series.

recrystallization. (1) The change from one crystal structure to another, such as occurs on heating or cooling through a critical temperature. (2)

The formation of a new, strain-free grain structure from that existing in cold-worked metal, usually accomplished by heating.

residual stress. Internal stress; stress present in a body that is free from external forces or thermal gradients.

robust design. An integrated system of tools and techniques that are aimed at reducing product or process performance variability while simultaneously guiding that performance toward an optimal setting. Robustness optimization is chiefly done for design concepts that are new so that the best values of the critical functional parameters are unknown. Robust design follows the methods first proposed by Genichi Taguchi.

root (of a notch). The innermost part of a *stress concentration,* such as the bottom of a thread or groove.

root cause analysis. Identification of the most basic or causal factor(s) that contribute to a failure.

root mean square (rms). A term describing the surface roughness of a machined surface, R_q, calculated as the square root of the average of the squared distance of the surface from the mean line. See also *surface roughness*.

rupture. Same as *fracture.*

S

selective leaching. Corrosion in which one element is preferentially removed from an alloy, leaving a residue (often porous) of the elements that are more resistant to the particular environment.

service loads. Forces encountered by a part or assembly during operation in service.

shear. A type of force that causes or tends to cause two regions of the same part or assembly to slide relative to each other in a direction parallel to their plane of contact. May be considered on a microscale when planes of atoms slide across each other during permanent, or plastic, deformation. May also be considered on a macroscale when gross movement occurs along one or more planes, as when a metal is cut or "sheared" by another metal.

shear fracture. Fracture that occurs when shear stresses exceed shear strength before any other type of fracture can occur. Typical shear fractures are transverse fracture of a ductile metal under a torsional (twisting) stress, and fracture of a rivet cut by sliding movement of the joined parts in opposite directions, like the action of a pair of scissors (shears).

shear lip. A narrow, slanting ridge, nominally about 45° to the surface, along the edge of a fracture surface where the fracture emerged from the interior of the metal. In the fracture of a ductile tensile specimen, the shear lip forms the typical "cup-and-cone" fracture. Shear lips may be present on the edges of some predominantly brittle fractures to form a "picture frame" around the surface of a rectangular part.

shear stress. See *stress.*

shot peening. A carefully controlled process of blasting a large number of hardened spherical or nearly spherical particles (shot) against the softer surface of a part. Each impingement of a shot makes a small indentation in the surface of the part, thereby inducing compressive residual stresses, which are usually intended to resist fatigue fracture or *stress-corrosion cracking.*

shrinkage cavity. A void left in cast metals as a result of solidification shrinkage, because the volume of metal decreases during cooling. Shrinkage cavities usually occur in the last metal to solidify after casting.

sintered metal. Same as powdered metal. Type of metal part made from a mass of metal particles that are pressed together to form a compact, then sintered (or heated for a prolonged time below the melting point) to bond the particles together.

slant fracture. A type of fracture appearance, typical of ductile fractures of flat sections, in which the plane of metal separation is inclined at an angle (usually about 45°) to the axis of the applied stress.

S-N curve. A plot of stress (S) against the number of cycles to failure (N). The stress can be the maximum stress (S_{max}) or the alternating stress amplitude (S_a). The stress values are usually nominal stress; that is, there is no adjustment for stress concentration. For S, a linear scale is used most often, but a log scale is sometimes used. Also know as *S-N diagram*.

solid metal embrittlement. The occurrence of embrittlement in a material below the melting point of the embrittling species. Also known as solid metal induced embrittlement. See also *liquid metal embrittlement.*

solution heat treatment. Heating an alloy to a suitable temperature, holding at that temperature long enough to cause one or more constituents to enter into solid solution, and then cooling rapidly enough to hold these constituents in solution.

spalling fatigue. See *subcase fatigue.*

spiral bevel gear. A type of bevel, or conical, gear in which the teeth are curved within the conical shape, rather than straight, as in a bevel gear. Compare with *hypoid.*

spline. A shaft with a series of longitudinal, straight projections that fit into slots in a mating part to transfer rotation to or from the shaft.

static. Stationary, or very slow. Frequently used in connection with routine tensile testing of metal specimens. Same as *monotonic.* Opposite of *dynamic,* or impact, testing or service.

strain. A measure of relative change in the size or shape of a body. "Linear strain" is change (increase or decrease) in a linear dimension. Usually expressed in inches per inch (in./in.), or millimeters per millimeter (mm/mm).

strength gradient. Shape of the strength curve within a part. The strength gradient can be determined by hardness tests made on a cross section of a part; hardness values are then converted into strength values, usually in pounds per square inch (psi), or megapascals (MPa).

stress. Force per unit area, often thought of as a force acting through a small area within a plane. It can be divided into components, perpendicular and parallel to the plane, called normal stress and shear stress, respectively. Usually expressed as pounds per square inch (psi), or megapascals (MPa).

stress concentration. Changes in contour, or discontinuities, that cause local increases in stress on a metal under load. Typical are sharp-cornered grooves, threads, fillets, holes, etc. Effect is most critical when the stress concentration is perpendicular (normal) to the principal tensile stress. Same as notch or stress raiser.

stress corrosion. Preferential attack of areas under stress in a corrosive environment, where such an environment alone would not have caused corrosion.

stress-corrosion cracking. Failure by cracking under combined action of corrosion and a tensile stress, either external (applied) or internal (residual). Cracking may be either intergranular or transgranular, depending on the metal and the corrosive medium.

stress cube. A finite volume of material used to depict three-dimensional states of stress and strain (displacement) distributions at a crack tip.

stress field. Stress distribution generated ahead of a sharp crack present in a loaded part or specimen. The stress field is characterized by a single parameter called stress-intensity, K.

stress-field analysis. Mathematical analysis of an assumed two-dimensional state of stress (plane-strain condition) at a crack tip in linear-elastic fracture mechanics.

stress gradient. Shape of a stress curve within a part when it is under load. In pure tension or compression, the stress gradient is uniform across the part, in the absence of stress concentrations. In pure torsion (twisting) or bending, the stress gradient is maximum at the surface and zero near the center, or neutral axis.

stress-intensity factor. A scaling factor, usually denoted by the symbol K, used in linear-elastic fracture mechanics to describe the intensification of applied stress at the tip of a crack of known size and shape. At the onset of rapid crack propagation in any structure containing a crack, the factor is called the critical stress-intensity factor, or the *fracture toughness*. Various subscripts are used to denote different loading conditions or fracture toughnesses:

K_c. Plane-stress fracture toughness. The value of stress sections thinner than those in which plane-strain conditions prevail.

K_I. Stress-intensity factor for a loading condition that displaces the crack faces in a direction normal to the crack plane (also known as the opening mode of deformation).

K_{Ic}. Plane-strain fracture toughness. The minimum value of K_c for any given material and condition, which is attained when rapid crack propagation in the opening mode is governed by plane-strain conditions.

K_{Id}. Dynamic fracture toughness. The fracture toughness determined under dynamic loading conditions; it is used as an approximation of K_{Ic} for very tough materials.

K_{ISCC}. Threshold stress intensity factor for stress-corrosion cracking. The critical plane-strain stress intensity at the onset of stress-corrosion cracking under specified conditions.

K_Q. Provisional value for plane-strain fracture toughness.

K_{th}. Threshold stress intensity for stress-corrosion cracking. The critical stress intensity at the onset of stress-corrosion cracking under specified conditions.

ΔK. The range of the stress-intensity factor during a fatigue cycle.

stress raiser. See *stress concentration*.

stress rupture strength. See *creep rupture strength*.

striations. Microscopic ridges or lines on a fatigue fracture that show the location of the tip of the fatigue crack at some point in time. They are locally perpendicular to the direction of growth of the fatigue crack. In ductile metals, the fatigue crack advances by one striation with each load application, assuming the load magnitude is great enough. Must not be confused with *beachmarks,* which are much larger and are formed in a different way.

stringers. In metals that have been hot worked, elongated patterns of impurities, or *inclusions,* that are aligned longitudinally. Commonly the term is associated with elongated oxide or sulfide inclusions in steel.

subcase fatigue. A type of fatigue cracking that originates below a hardened *case,* or in the *core.* Large pieces of metal may be removed from the surface because of very high compressive stresses, usually on gear teeth. Also called spalling fatigue and case crushing.

superalloys. Nickel-, iron-nickel-, and cobalt-base alloys generally used at temperatures above approximately 540 °C (1000 °F). They exhibit a combination of mechanical strength and resistance to surface degradation that make them suitable for applications requiring high heat and corrosion resistance.

surface roughness. Fine irregularities in the surface texture of a material, usually including those resulting from the inherent action of the production process. Surface roughness is usually reported as the arithmetic roughness average, R_a, and is given in micrometers or microinches.

T

Taguchi methods. See *robust design*.
tangential. See *circumferential*.
technical plan for resolution (TPR) chart. A spreadsheet used in a failure investigation to plan the detail process for proving or disproving

the potential root causes of failure. A TPR chart is developed from and used in conjunction with a *failure mode assessment (FMA) chart.*

tensile. Pertaining to forces on a body that tend to stretch, or elongate, the body. A rope or wire under load is subject to tensile forces.

tensile strength. In tensile testing, the ratio of maximum load to the original cross-sectional area.

thermal cycles. Repetitive changes in temperature, that is, from a low temperature to a higher temperature, and back again.

through hardening. Hardening of a metal part, usually steel, in which the hardness across a section of the part is essentially uniform; that is, the center of the section is only slightly lower in hardness than the surface.

tolerance. The specified permissible deviation from a specified nominal dimension, or the permissible variation in size or other quality characteristic of a part.

torque. A measure of the twisting moment applied to a part under a torsional stress. Usually expressed in terms of inch pounds or foot pounds, although the terms "pound inches" and "pound feet" are technically more accurate for torsional moments.

torsion. A twisting action applied to a generally shaft-like, cylindrical, or tubular member. The twisting may be either reversed (back and forth) or unidirectional (one way).

total quality management (TQM). A management approach to long-term success that emphasizes customer satisfaction. The elements of TQM are continuous improvement, involving all members of the organization, and decision making based on data, using a set of simple TQM analysis tools.

toughness. Ability of a material to absorb energy and deform plastically before fracturing. Toughness is proportional to the area under the stress-strain curve from the origin to the breaking point. In metals, toughness is usually measured by the energy absorbed in a notch impact test.

transgranular fracture. Through, or across, the crystals or grains of a metal. Same as transcrystalline and intracrystalline. Contrast with *intergranular fracture.* The most common types of transgranular fracture are fatigue fractures, *cleavage fractures, dimpled rupture fractures,* and *shear fractures.*

transverse. Literally "across," usually signifying a direction or plane perpendicular to the axis of a part.

U

underbead crack. A subsurface crack in the base metal near a weld.
undercut. In welding, a groove melted into the base metal adjacent to the toe, or edge, of a weld and left unfilled.

uniform corrosion. (1) A type of corrosion attack (deterioration) uniformly distributed over a metal surface. (2) Corrosion that proceeds at approximately the same rate over a metal surface. Also called general corrosion.

W

wear. The undesired removal of material from contacting surfaces by mechanical action.
worm gear. A type of gear in which the gear teeth are wrapped around the shaft-like hub, somewhat as threads are wrapped around a bolt or screw.

Y

yield point. The first stress in a material, less than the maximum attainable stress, at which an increase in strain occurs without an increase in stress. Not a general term or property; only certain metals exhibit a yield point.
yield strength. The stress at which a material exhibits a specified deviation from proportionality of stress and strain. The specified deviation is usually 0.2% for most metals. A general term or property, preferred to *yield point.*
Young's modulus. Same as *modulus of elasticity.*

SELECTED REFERENCES

- *ASM Materials Engineering Dictionary,* J.R. Davis, Ed., ASM International, 1992
- *Failure Analysis and Prevention,* Vol 11, *ASM Handbook,* ASM International, 2002
- *Heat Treating,* Vol 4, *ASM Handbook,* ASM International, 1991
- *Materials Selection and Design,* Vol 20, *ASM Handbook,* ASM International, 1997
- *Metals Handbook Desk Edition,* 2nd ed., ASM International, 1998
- D. Wulpi, *Understanding How Components Fail,* 2nd ed., ASM International, 1999

APPENDIX 3

Suggested Further Reading

- J. Berk and S. Berk, *Total Quality Management,* Sterling, 1993
- C.R. Brooks and A. Choudhury, *Metallurgical Failure Analysis,* McGraw-Hill, 1993
- V.J. Colangelo and F.A. Heiser, *Analysis of Metallurgical Failures,* 2nd ed., John Wiley & Sons, 1987
- D.P. Dennies, "The Organization of a Failure Investigation," *Practical Failure Analysis,* June 2002
- *Failure Analysis and Prevention,* Vol 11, *ASM Handbook,* ASM International, 2002
- *Journal of Failure Analysis and Prevention* (formerly *Practical Failure Analysis*), ASM International, published bimonthly
- C.H. Kepner and B.B. Tregoe, *The New Rational Manager,* Kepner-Tregoe, 1997
- R.J. Latino and K.C. Latino, *Root Cause Analysis: Improving Performance for Bottom-Line Results,* 2nd ed., CRC Press, 2002
- C.R. Nelms, *What You Can Learn from Things That Go Wrong, a Guide Book to the Root Causes of Failure,* Failsafe Network, 1994
- C.R. Nelms, *The Dynamics of Inculcating the Root Cause Mentality,* Failsafe Network, 1995
- *Principles of Failure Analysis,* 15-lesson self-study course, ASM International, revised 2002
- D.A. Ryder, *The Elements of Fractography,* AGARD-AG-155-71, 1971
- N. Schlager, Ed., *When Technology Fails: Significant Technological Disasters, Accidents, and Failures of the 20th Century,* Gale Group, 1994
- J.J. Scutti and W.J. McBrine, Introduction to Failure Analysis and Prevention, *Failure Analysis and Prevention,* Vol 11, *ASM Handbook,* ASM International, 2002, p 3–23

- G.F. Vander Voort, "Conducting the Failure Examination," *Practical Failure Analysis,* April 2001
- C.R. Walker and K.K. Starr, *Failure Analysis Handbook,* Pratt & Whitney, WRC-TR-89-4060, 1989
- D.J. Wulpi, *Understanding How Components Fail,* 2nd ed., ASM International, 1999

Index

A

Acetate tape replicas 161, 193, 194
Aging 33–34, 169, 176, 186
Aircraft/aerospace industry 32–33. *See also Columbia* Space Shuttle; International Space Station (ISS)
  AISI 440C stainless steel, 32
  Apollo 1, 1–2
  bird strikes, 40
  British Overseas Airways Corporation (BOAC) Comet, 20–25
  failure statistics, 61(T), 62, 63(T), 64(T), 65(F), 66–67
  fastener lubrication, 51
  7050 aluminum alloy, 68–69
  Soviet rocket fire, 121–124
  VLS-1 rocket accident, 125
  welding of dissimilar titanium alloys, 50
AISI 52100 32
Aluminum
  high-strength, 179, 180, 181
  melting temperature, 31, 46
  7050 alloy, 65, 68–69, 135
  superpure, 182
Analysis. *See* Testing/analysis
Andrea Doria-Stockholm Collision 144–146, 147(F), 148–149
Applied stress 182
Assembly errors 36, 38(F), 41, 42(F), 43–45
ASTM specifications 107, 154, 188
Atomic-absorption spectroscopy 187
Auger electron spectrometer 187
Austenitic stainless steel 52, 176, 179, 180(F)
  prior-austenite grain boundaries, 168(F), 181, 182(F)
Automobile industry 17–18, 44–45, 49–50

B

Band-aid fixes 131, 136
Biotech implants 35
Boeing Company 24, 25, 41
Borescopes 103
British Overseas Airways Corporation (BOAC) Comet 20–25

C

California energy crisis 12–13
Cameras 97, 99(F), 100(T), 101
Case depth 172
Cavitation 9, 10(F), 183
Cellular phones 99(F), 100(T), 102
Charpy test 160
Chemical analysis 185, 186
Chemical milling 15, 16(F)
Chrome plating process 32–33
Color charts 100(T), 101
Columbia Accident Investigation Board (CAIB) 18–19
Columbia **Space Shuttle** 12, 18–19
  (SSME) high-pressure oxidizer turbopump, 42(F), 43, 46–47
Combing Process 143
Component failure 34–35
Conductivity tester 100(T), 102
Corrective action assessment (CAA) chart 118, 119(F), 120
Corrective Action Review Plan 81
Corrective actions 118, 119(F), 120, 127–136
Corrosion 20, 29–30, 32, 34, 39, 63(T), 64(T), 65(F), 67, 85, 88, 89, 161
Cracking. *See also* Fractures
  branching, 178, 179, 180(F), 182
  growth, 155, 164, 176, 178, 181, 182(F)
  hairline, 181–182
  hydrogen assisted growth, 181, 182(F)
  propagation, 175, 185(F)
  secondary, 162–163
Crack-opening displacement test 188
Crankshafts 14, 15(F)
Creep and stress rupture 182, 183–184(F), 185

Crystallization 24
Custom 455 CRES 50–52
Cyclic stress 173, 177, 180

D

Database 61(T), 62(T), 63(T), 64(T), 65(F), 66–67, 71
Data collection 62(T), 63(T), 64(T)
Decompression, explosive 22, 23, 24, 25
Delta ferrite 99(F)
Depressurization 21, 24
Design errors 36, 49–52
Distortion 67, 85, 108, 159
Ductility 160, 174, 175, 178, 181
Dye penetrant inspection 39–40, 68
Dynamic tear test 188

E

Electrical discharge machining (EDM) 100(T), 102
Electron-microprobe (SEM-EDX) 181, 187
Energy-dispersive spectroscopy (EDS) 177
Energy-dispersive x-ray (EDX) 187
Equipment. *See* Tools
Evidence freezing 106–112
Extra-low interstitial (ELI) titanium alloy 134
Eyewitness statements 100(T), 102, 103–107

F

Fabrication/manufacturing errors 45–49
Failsafe Network, Inc. 5, 107, 142–143
Failure analysis report 192
Failure analysts 2–4
  attitude, 99, 100(T)
  communication skills, 137–139
  help, asking for, 128, 129, 132
  integrity, 139–140, 194
  legal ramifications, 6–7, 194
  skills, 137
  understanding the problem, 128–129, 143
Failure cause 3–4. *See also* Root cause
  abnormal service damage, 62(T), 63(T)
  human factors, 5, 30–31, 52
  industry specific, 32–34
  jumping to conclusions, 129
  levels of, 5(F), 6, 30–31
  prevention of, 4
  recognition of, 7–10(F)
  relative importance of, 65(F), 69(T), 70(T)
  unique nature of, 31–32

Failure investigation 3. *See also* Stages of failure analysis
  analytical techniques, 186–188
  benefits of, 71
  foregoing, 134
  nine steps of, 91
  on-site, 193(F), 194
  pitfalls, 128–131
  planning for, 77–78
  reasons for, 67–71
  substitutes for, 131–136
  of surfaces and deposits, 187–188
  team composition, 104
  timeline, 78(F), 79, 80(F), 130
  understanding the failure, 91, 96–97, 98(F), 99(F), 130
Failure mechanism 5, 30–31, 46
Failure mode and effects analysis (FMEA) 143
Failure mode assessment (FMA) chart 113, 114(F), 115
Failure modes 168(F), 173–177, 178(F), 179, 180(F), 181, 182–184(F), 185
Failure prevention 4–6
Fatigue 22, 23, 30
  crack, 24, 25, 29, 170, 174, 178, 185
  discoloration/crystallization, 24
  high-cycle, 67, 131
  life, 33, 34
  repetitive load fracture, 177, 178(F), 179
  stoppers, 25
  striations, 56, 57(F), 168(F), 177, 179, 180, 183
Fault tree analysis (FTA) 77, 80(F), 81, 101, 109(F), 110–113, 115, 118, 119(F)
Federal Aviation Administration (FAA) 21, 40
Federal Rules of Evidence 107, 151
Field investigation kit 97, 98(F), 99(F), 100(T), 101–103
Finite-element analysis (FEA) 173, 189
Firestone tire failure 16–18, 26, 82, 83(F), 84–87
Flashlight 100(T), 102
Forms 100(T), 104–105, 107. *See also* Eyewitness statements
Fractography 175, 183, 184(F), 185
Fractures. *See also* Cracking; Fatigue; Stages of failure analysis
  brittle, 164(F), 171, 175
  chevron marks, 155, 164(F), 165, 176
  cleaning, 161–162
  cleavage, 167(F), 175–176
  crystalline, 176
  dimpled, 167(F), 185
  ductile, 174, 174–175, 175
  intergranular, 168(F), 177, 182(F)
  mechanics, 188
  piston rod, 170(F)

stress-corrosion cracking (SCC), 29, 34, 169, 174, 179, 180(F), 181
sectioning, 162
sequence establishment, 155, 156(F)
shear lip, 164, 165(F), 170(F), 175, 176
study of, 157
toughness, 30, 33, 188
Fracture toughness test 188
Fretting 178
F-20 Tigershark 10(F)

G

Gas chromatography 187
Gears 15(F)
Grain boundaries 166, 167–168(F), 169, 175, 177, 181, 182(F), 183, 184(F)

H

Hardness testing 30, 159, 194
Heat treatments
  annealing cycle, 31
  decarburization, 169, 171
  hardness testing, 30
  hydrogen removal, 32, 33
  interpretation/knowledge of, 48
  precipitation-aging, 33
  solution, 32, 48–49, 51
  steel specifications, 48
Hot isostatic processing (HIP) 136
House Subcommittee on Oversight and Investigations 83(F)
Hydrogen-assisted cracking 168(F), 181, 182(F)
Hydrogen embrittlement 165, 168(F), 179, 181, 182, 186
Hydrogen removal 32–33

I

Infrared spectroscopy 187
Intergranular stress corrosion cracking (IGSCC) 167, 168(F)
International Space Station (ISS)
  Custom 455 CRES, 48
  extravehicular activities (EVAs), 11(F), 12
  heat treatments, 48
  welding of dissimilar titanium alloys, 50
Interviews 103
Inventory 155
Ion-microprobe analyzer 187
Izod specimens 160

K

Kepner-Tregoe (KT) method 142
Khrushchev, Nikita S. 121, 122, 125

L

Laptop/notebook computers 99(F), 100(T), 101
Leatherman tool 100(T), 102
Liquid metal embrittlement (LME) 46, 51, 181
Liquid-penetrant inspection 158
Loupes 102

M

Machine drawings 52
Magnetic-particle inspection 158
Magnets 100(T), 101–102
Magnifying glasses 100(T), 102
Maintenance improprieties 36–39
Manufacturing errors. *See* Fabrication/manufacturing errors
Markers, indelible ink 100(T), 102
Martensite 52, 99, 171, 181
Material selection 61(T), 62(T)
Mechanical testing 159–160
Measuring devices 101
Metallographic examination 169, 170–171(F), 172
Metallography unit 193
Metallurgical analysis 31–32
Metallurgical laboratories 29
Mirrors 102

N

NACE International 39
Nacelles 38(F), 43, 44
NASA 11, 12, 18, 19
National Highway Traffic Safety Administration (NHTSA) 84–87
National Transportation Safety Board (NTSB) 2
Nondestructive testing (NDT) 158–159, 163, 189
Notch effects 188

O

Oil field industry 33
Operational temperature 52

Operation Failsafe 143
Overload failure 173, 174

P

Pareto diagram 64(F)
Personal digital assistant (PDA) 101, 102
Photographic records 154
Pitting 67, 171, 172(F), 179
Plastic deformation 157, 163, 182, 183
Polycrystalline specimens 176
Portable laboratories 193(F), 194
Precipitation-hardening (PH) martensitic steel 48, 51, 52, 99(F), 106
Pressurization 21–24, 25(F)
PROACT software 142–143
Probability of detection (POD) 134
Problem solving. *See* Puzzles/problem solving
Process failure 34
Product disassembly 130–131
Puzzles/problem solving 71, 72–76(F)
  four-step problem-solving process, 76(F)

R

Radiography 159
Recrystallization 180
Republic Steel Corporation 36
Residual stress analysis 159
Rockwell hardness level 30, 159
Rollovers. *See* Sport utility vehicle (SUV) rollovers
Room-temperature-vulcanizing (RTV) rubber replicas 193, 194
Root cause 2, 5, 29, 30–31
  brainstorming, 110–111
  cause-and-effect, 112, 115
  converging on, 91, 115, 116(F), 117–118, 120
  determining, 61(T), 68, 77
  evaluating, 91, 113, 115, 117, 120
  identifying, 91, 108–112, 129
  prioritizing, 113
  sequence of testing/analysis, 117
Root cause analysis (RCA) investigation process 142
Root Cause Conference 143
ROOTS investigative Process 143
Royal Aircraft Establishment (RAE) 22–24
Rupture. *See* Fracture

S

Safety of Life at Sea (SOLAS) 149
Sample collection 154–155

Scanning electron microscope (SEM) 161, 194
Scene preservation 97, 98(F), 99, 103, 105–107
Scraping tools 102, 103
Secondary mass spectrometer (SIMS technique) 187
Secrecy, failures 124–125
Securities and Exchange Commission (SEC) 87
Self-excitation 57–58
Semiquantitive emission spectrography 187
Service history 153–154
Shadowing 48
Shainin 143
Significant 6 Barriers to Root Cause Discovery 143
Silver bullet theory 128
Simulated-service testing 189
Situation-Filter-Outcome Model 143
Soviet rocket fire 121–124
Space Shuttle. *See* Columbia Space Shuttle
Specification failure 35–36
Spectrography 187
Spectrophotometry 187
Sport utility vehicle (SUV) rollovers 16. *See also* Firestone tire failure
  driving habits of owners, 17
  government safety laws for, 18
Spot testing 188
Stages of failure analysis. *See also* Failure modes
  chemical analysis, 185, 186
  collection of background data/samples, 153–155, 156(F), 157
  conclusions/reports, 190–192
  consulting, 189–190
  follow-up, 192–193
  fracture mechanics, 188
  macroscopic examination, 163, 164–165(F)
  mechanical testing, 159–160
  metallographic examination, 169, 170–171(F), 172
  microscopic examination, 166–168(F), 169
  nondestructive inspection, 158–159
  preliminary examination, 157–158
  selection/preservation, 160–163
  stress analysis, 172(F), 173
Statistics 61(T), 62(T), 63(T), 64–65. *See also* Database
Strain 173, 176, 183, 186, 188
Stress
  concentration, 26, 174, 177, 183, 188
  relaxation, 183
  rupture failure, 30, 64, 65(F), 177, 183
  tensile, 164, 169, 175, 176, 177, 179, 183
Stress-corrosion cracking (SCC) 29, 34, 169, 174, 179, 180(F), 181, 185(F)

Superalloys
  cobalt-base, 31, 48
  ductility, 183
  iron-base heat resistant, 32, 130
  nickel-base, 33, 171
  nitrogen levels, 186
Surface finish comparators 100(T), 102
Suspended laser acquisition pods (SLAPs) 35

T

Tacoma Narrows Bridge Collapse 53–54, 55(F), 56, 57(F), 58
Taguchi analysis 66
Technical plan for resolution (TPR) chart 115, 116(F), 117–118
Tensile tests 30, 48, 49, 160
Testing/analysis. *See also individual listings*
  checklist, 191
  chemical, 185–186
  "design of experiments," 65(F), 66
  errors, 40–41
  hardness, 30, 159, 194
  mechanical, 159–160
  nondestructive (NDT), 158–159, 163, 189
  runaway, 129
  sequence, 117
  surfaces and deposits, 187–188
  techniques, 186–188
  for toughness, 188
Thick-lip stress rupture 184(F)
Tire Industry Safety Council 86
Titanic 93–96
Titanium
  chemical milling, 15, 16(F)
  commercially pure (CP), 50
  corrosion resistance, 89
  domes, 15, 16(F)
  ELI grade, 135
  hot isostatic pressing, 136
  liquid metal embrittlement, 51
Tools 142–143, 194
Torsional forces 58(F), 172, 174
Total Quality Management and Continuous Improvement 71

Transgranular cleavage 175–177
Triaxiality 188
TWA Flight 800 2, 70(T), 71

U

Ultraviolet spectroscopy 187
Ultrasonic inspection 158–159
Uphill diffusion mechanism 50
U.S. Civil Aeronautics Administration (CAA) 21
U.S. Department of Transportation 39
USS *Greeneville* 13(F), 14
USS *Ogden* 14

V

Vacuum furnaces 48–49
Visual inspection 157

W

Wavelength-dispersive x-ray (WDX) 187
Welding 46–50
Wet chemical analysis 187
Whitestone Bridge 56, 57(F)
WHY Tree 143
Wind engineering 58
Wind-tunnel testing 57, 58, 59, 189

X

X-ray diffraction 187
X-ray fluorescence spectrographic technique 187

Y

Yield strength 177, 179